Water Reuse Policies for Potable Use

As water demand has increased globally and resources have become more limited because of physical scarcity, over-exploitation and pollution, it has been necessary to develop more options for water supplies. These options include the production at large scale of high-quality reused water from municipal sources for potable uses. Their economic, social and environmental benefits have been many as they have addressed supply scarcity, efficient resource use and environmental and public health considerations.

This book includes discussions on potable water reuse history; emerging contaminants and public health; public-private partnerships in the water reuse sector; regulatory frameworks for reused water in the United States and Europe; experiences in Australia, China in general and Beijing in particular, Singapore and Windhoek; narratives and public acceptance and perceptions of alternative water sources.

The main constraints on implementation of water reuse projects in different parts of the world seem to have been lack of full public support due to perceived health hazards and environmental impacts. A main handicap has been that governments and water utilities have been slow to understand public concerns and perceptions. After several backlashes, public information, communication and awareness campaigns, broader participation and educational programmes have become integral parts of development policy and decision-making frameworks.

Cecilia Tortajada is a Senior Research Fellow at the Institute of Water Policy, Lee Kuan Yew School of Public Policy, National University of Singapore, Singapore.

Choon Nam Ong is the Director of the NUS Environmental Research Institute (NERI) and Professor at the Saw Swee Hock School of Public Health, both at National University of Singapore, Singapore.

Routledge Special Issues on Water Policy and Governance

https://www.routledge.com/series/WATER

Edited by
Cecilia Tortajada (*IJWRD*), *Institute of Water Policy, Lee Kuan Yew School of Public Policy, National University of Singapore, Singapore*
James Nickum (*WI*), *International Water Resources Association, France*

Most of the world's water problems, and their solutions, are directly related to policies and governance, both specific to water and in general. Two of the world's leading journals in this area, the *International Journal of Water Resources Development* and *Water International* (the official journal of the International Water Resources Association), contribute to this special issues series, aimed at disseminating new knowledge on the policy and governance of water resources to a very broad and diverse readership all over the world. The series should be of direct interest to all policy makers, professionals and lay readers concerned with obtaining the latest perspectives on addressing the world's many water issues.

Water Reuse Policies for Potable Use

Edited by
Cecilia Tortajada and Choon Nam Ong

Water for All: Conserve, Value, Enjoy

Routledge
Taylor & Francis Group
LONDON AND NEW YORK

Lee Kuan Yew
School of Public Policy
National University of Singapore

First published 2017
by Routledge

2 Park Square, Milton Park, Abingdon, Oxfordshire OX14 4RN
52 Vanderbilt Avenue, New York, NY 10017

Routledge is an imprint of the Taylor & Francis Group, an informa business

First issued in paperback 2018

British Library Cataloguing in Publication Data
A catalogue record for this book is available from the British Library

ISBN 13: 978-1-138-63728-3 (hbk)
ISBN 13: 978-0-367-10954-7 (pbk)

Typeset in MyriadPro
by diacriTech, Chennai

Publisher's Note
The publisher accepts responsibility for any inconsistencies that may have arisen
during the conversion of this book from journal articles to book chapters, namely
the possible inclusion of journal terminology.

Disclaimer
Every effort has been made to contact copyright holders for their permission to
reprint material in this book. The publishers would be grateful to hear from any
copyright holder who is not here acknowledged and will undertake to rectify any
errors or omissions in future editions of this book.

Contents

CONTENTS

Citation Information

The chapters in this book were originally published in the *International Journal of Water Resources Development*, volume 32, issue 4 (July 2016). When citing this material, please use the original page numbering for each article, as follows:

Chapter 3

Public–private partnerships in the water reuse sector: a global assessment
David A. Lloyd Owen
International Journal of Water Resources Development, volume 32, issue 4 (July 2016)
pp. 526–535

Chapter 4

The regulatory framework of reclaimed wastewater for potable reuse in the United States
Rosario Sanchez-Flores, Adam Conner and Ronald A. Kaiser
International Journal of Water Resources Development, volume 32, issue 4 (July 2016)
pp. 536–558

Chapter 5

Common or independent? The debate over regulations and standards for water reuse in Europe
John Fawell, Kristell Le Corre and Paul Jeffrey
International Journal of Water Resources Development, volume 32, issue 4 (July 2016)
pp. 559–572

Chapter 6

Policy issues confronting Australian urban water reuse
James Horne
International Journal of Water Resources Development, volume 32, issue 4 (July 2016)
pp. 573–589

Chapter 7

Wastewater reuse in Beijing: an evolving hybrid system
Olivia Jensen and Xudong Yu
International Journal of Water Resources Development, volume 32, issue 4 (July 2016)
pp. 590–610

Chapter 8

Singapore's experience with reclaimed water: NEWater
Hannah Lee and Thai Pin Tan
International Journal of Water Resources Development, volume 32, issue 4 (July 2016)
pp. 611–621

Chapter 9

Overcoming global water reuse barriers: the Windhoek experience
P. van Rensburg
International Journal of Water Resources Development, volume 32, issue 4 (July 2016)
pp. 622–636

Chapter 10

A lived-experience investigation of narratives: recycled drinking water
Leong Ching
International Journal of Water Resources Development, volume 32, issue 4 (July 2016)
pp. 637–649

Chapter 11

*Public acceptance and perceptions of alternative water sources: a comparative study
in nine locations*
Anna Hurlimann and Sara Dolnicar
International Journal of Water Resources Development, volume 32, issue 4 (July 2016)
pp. 650–673

For any permission-related enquiries please visit:
http://www.tandfonline.com/page/help/permissions

Notes on Contributors

Asit K. Biswas is the Founder of the Third World Centre for Water Management in Mexico, and currently is a Distinguished Visiting Professor at the Lee Kuan Yew School of Public Policy, National University of Singapore, SIngapore.

Leong Ching is the Deputy Director, Institute of Water Policy, Lee Kuan Yew School of Public Policy, National University of Singapore, Singapore, and Assistant Professor at the same school.

Adam Conner is a Water Planner at Water Resources, San Antonio Water System, Texas, USA.

Joseph A. Cotruvo is the Founder of Joseph Cotruvo & Associates LLC, Washington, USA.

Sara Dolnicar is a Professor at UQ Business School, The University of Queensland, Brisbane, Australia.

John Fawell is a Faculty Member at Cranfield Water Science Institute, Cranfield University, UK.

James Horne is a Visiting Fellow at the College of Asia and the Pacific, Australian National University, Canberra, Australia.

Anna Hurlimann is a Senior Lecturer at the Faculty of Architecture, Building and Planning, The University of Melbourne, Australia.

Paul Jeffrey is Professor at Cranfield Water Science Institute, Cranfield University, UK.

Olivia Jensen is a Senior Research Fellow, Institute of Water Policy, Lee Kuan Yew School of Public Policy, National University of Singapore, Singapore.

Ronald A. Kaiser is a Professor and the Chair of the Texas A&M University Water Program, Texas A&M University, USA.

Kristell Le Corre is Research Fellow in Water Reuse at the Cranfield Water Science Institute, Cranfield University, UK.

Hannah Lee formerly at PUB The National Water Agency, Singapore.

David A. Lloyd Owen is the Founder and Managing Director of Envisager Limited, Ceredigion, UK.

Peter Joo Hee Ng is Chief Executive of PUB The National Water Agency, Singapore.

Choon Nam Ong is the Director of the NUS Environmental Research Institute (NERI) and a Professor at the Saw Swee Hock School of Public Health, both at National University of Singapore, Singapore.

P. van Rensburg is a Strategic Executive for Infrastructure, Water and Technical Services, City of Windhoek, Namibia.

Rosario Sanchez-Flores is a Research Scientist for the Water Management and Hydrological Sciences, Texas A&M University, College Station, USA.

Thai Pin Tan is the Director (Water Supply), PUB The National Water Agency, Singapore.

Cecilia Tortajada is a Senior Research Fellow at the Institute of Water Policy, Lee Kuan Yew School of Public Policy, National University of Singapore, Singapore.

Xudong Yu is a PhD student at the School of Environment and Natural Resources, Renmin University of China, Beijing.

Foreword

Peter Joo Hee Ng

PUB (Singapore's National Water Agency)

By now it is quite clear that the world is plunging headlong into water crisis. The warnings have been loud, and they are dire. Some of the latest come from the World Economic Forum (2015), which has unequivocally declared water – the scarcity of, the lack of access to, and the poisoning of – to be the biggest threat to human life in the next decade.

Water and the security of its supply have preoccupied Singapore's leaders and decision makers since the city-state's independence half a century ago. The late Lee Kuan Yew, Singapore's first prime minister, recognized from day one that enduring water security was nothing less than an existential challenge. Mr Lee had devoted his entire political life to securing Singapore's water future and once famously recalled, "Water dominated every other policy. Every other policy had to bend at the knees for water survival." So, perhaps more than any other country, Singapore has always treated the possibility that there would not be enough water as neither novel nor remote.

Unsurprisingly, then, reuse is a plank of Singapore's water strategy. At PUB – Singapore's national water agency – sewage treatment works are called *water reclamation plants*. Because, in our minds, the H_2O molecule is never lost and water is an endlessly reusable resource. Water can always be reclaimed and retreated so that it can be drunk again. PUB leads the world in this, and today we are able to, literally, turn wastewater into sweet water for very little money. PUB reclaims every drop of sewage and, for more than a decade, has turned much of it into drinking water again.

The Industrial Revolution gave humanity machines, factories and mass production, and greatly increased incomes and the standard of living. It also gave us an economy that takes, makes and disposes, generating massive amounts of waste in the process. Now, of course, we readily admit and acknowledge that taking the earth's resources to make the things that we desire, and then throwing them away when we do not want them anymore, is just not a sustainable way of doing things. Indeed, when it comes to life-giving water, it is simply unacceptable that it should be discarded after just one use.

Very early on, Singapore invested great effort and resources into researching technologies that would make wastewater potable again. When these become viable at the turn of new millennium, PUB adopted them with great zeal and promptly started manufacturing NEWater, which is ultra-high-quality recycled water, on an industrial scale. Today, we have enough NEWater capacity to supply about 40% of Singapore's daily demand.

Wastewater reuse is particularly attractive to Singapore because it is a drought-resistant source of potable water. The requisite treatment technologies that are involved have become commonplace; their reliability and efficacy are well-established and still improving by the day. Even better, and unknown to most, making sewage potable actually requires, litre for litre, far less energy than desalinating seawater.

Of course, the challenge, even for us in Singapore, is to persuade people to imbibe it after we have made the stuff. This remains a tricky issue the world over, as the various contributors to this volume make amply clear. Like it or not, the average person, if free to choose, will shun drinking processed wastewater, even if he believes it to be perfectly safe.

Public acceptance of NEWater has been high in Singapore, bolstered in part by the country's pre-existing water-stressed conditions. But we take nothing for granted, and continue to retain a careful and cautious approach. Thus, even though NEWater is entirely potable, we have not yet rushed into direct potable reuse.

When it comes to direct potable reuse, public acceptance and good regulation are two sides of the same coin. One reinforces the other. My own view is that, driven by necessity, direct potable reuse will come sooner rather than later. And it will become widely practised and a new norm once enough reputable jurisdictions enact suitable regulation.

Every Singaporean grade schooler is taught the hydrologic cycle and knows how Mother Nature reclaims and recycles water in all its forms. What we do in PUB's water reclamation plants and NEWater factories is, in essence, copying nature's way. In Singapore, we have every motivation to do this. I suspect the rest of the world will increasingly have to do the same.

Reference

World Economic Forum. (2015). Outlook on the Global Agenda 2015. Retrieved from http://reports. weforum.org/outlook-global-agenda-2015/wp-content/blogs.dir/59/mp/files/pages/files/outlook-2015-a4-downloadable.pdf

Preface

Over the past decade, the issue of water crisis has become increasingly popular. In early April 2016, if one put 'water crisis' in Google, there were some 30 million references. Irrespective of where in the world one is located, very seldom does a week pass without a major story in the media on the so-called water crisis.

Any serious objective and comprehensive analysis of the overall water crisis issue will indicate that the world is not facing a crisis because of actual physical scarcities of water. This is true even for the most arid inhabited regions of the world. However, the world is indeed facing a crisis, because of continued mismanagement of water over decades, if not centuries. If the management process can be improved, there will be enough water in the world for all types of water uses, not only for the present but also for 2050. By 2050, the world population is projected to reach 9.7 billion, that is, 2.3 billion more than at present. Furthermore, human activities are expected to increase very significantly during this period. But even by 2050, by using good management practices, knowledge and technology available at present (in contrast to what may be available in the coming decades), and formulation and implementation of rational water policies, water crisis can be avoided. Thus, as Shakespeare said many centuries ago, the fault is with us and "not in our stars".

A major problem for the water sector has been the lack of realization that water is a renewable resource. It is not like oil or coal, which once used breaks down into various components and cannot be used again. Water is a renewable resource. It can be used, treated and then reused. This cycle can be repeated numerous times with good management.

For the domestic water use sector, there is currently enough knowledge and technology which can be successfully used to treat wastewater to a level that it can be of even better quality than the water supplied by the utilities. Singapore at present treats its wastewater so well that it is of better quality than the drinking water that its citizens receive. The treatment process is cost-effective. Monitoring and supervision of the treatment processes are very rigorous and strict so that water quality requirements are met consistently. There are no health risks, real or perceived. This means that in terms of technology and economics, there is absolutely no reason why all wastewater generated is not being treated extensively at present so that it can be reused time and again.

While the future is mostly difficult or even impossible to foresee, one prediction can be made with complete certainty. Scientific and technological advances in the coming decades will make wastewater treatment processes continually more and more efficient and cost-effective. In future, it would be a crime not to treat wastewater properly and then reuse it as many times as possible.

The legitimate question thus is, if wastewater can be cost-effectively treated to the level of drinking water, or even better, at present, why is this practice not being used extensively all over the world?

The reasons are many. However, the two most important ones are the following. First, historically, with the exception of very few countries like Singapore, most unfortunately, water has not been high on the national political agendas. Politicians and media are interested in water only when there are catastrophes like heavy floods, prolonged droughts or other serious national disasters like earthquakes. Once these catastrophes are over, political and media interests in water simply evaporate. Unfortunately, water problems can be solved only on a long-term basis, which requires sustained political interest. This is missing at present.

In addition, to the extent there is interest, this is almost exclusively with respect to water quantity; quality issues are seldom seriously considered. There is lot of rhetoric on the importance of water quality from national governments and international organizations. Sadly, sustained and well-thought-out actions are conspicuous by their absence. Thus, not surprisingly, almost all water bodies in and around centres of population and industrial developments in developing countries are heavily polluted with known and unknown contaminants. Furthermore, in most cases, water quality is steadily deteriorating, in both surface water bodies and aquifers.

Second, the fundamental problem with the use of properly treated wastewater is neither technological nor economical, but public acceptance of its use. No matter how well wastewater is treated, people remember its history, and are strongly opposed to its use even though its quality could be higher than provided by a utility. Even the discussions of the issue are often framed in pejorative terms. Thus, in California, the discussions were framed on the concept of "toilets to taps", and in Australia, it became "citizens against drinking sewage".

Generally people are uncomfortable with or averse to the idea of drinking treated waste-water, irrespective of all the scientific and technical evidence that categorically indicates that it is perfectly safe to drink. One study in the United States found that one in four people believe that wastewater cannot be treated to a high enough quality so that it can be actually drunk. In other words, no matter what is done, no matter what its quality is, its history ensures that its use is unacceptable. There appears to be a universal belief that once water has been in contact with a disgusting object like human excreta, it will always remain in contact, irrespective of the treatment processes.

There are other anomalies in human perception as well. One American study indicated that people are more willing to drink treated wastewater if it has been stored in an aquifer for 10 years, compared to only 1 year. Some 40% of the people are willing to try it if it has travelled in a river for a hundred miles, as opposed to only one mile. Somehow, it seems, the longer it is stored, or the further it travels, more acceptable it becomes.

In other words, irrespective of all the scientific and technical evidence which indicates that treated wastewater is safe to drink, the gut feeling of most citizens is that it is not so, primarily because of the 'yuck factor'.

This leads to another issue. Historically, the water profession is dominated by engineers and technologists. There is no question that in the coming years they will make the treatment processes continually more efficient and progressively more cost-effective. However, this development alone is unlikely to change the opinions of the vast majority of the people in terms of drinking properly treated wastewater. The breakthroughs are likely to come only if behavioural psychologists and economists play a major role in convincing the society that their attitudes to and perceptions of use of treated wastewater are irrational. This behaviour needs to undergo a sea change, which will be difficult to achieve without the help of behavioural scientists.

The good news is that behavioural psychologists and economists are starting to take an interest in this issue. While this is a good beginning, we need a critical mass of such experts working all over the world to convince the people that current practices of wastewater treatment can ensure that the resulting water is perfectly safe to drink. This has to be complemented with proper long-term monitoring and safeguard practices.

If water is not reused extensively, the world's water problems cannot be solved. It is thus very heartening to see that National University of Singapore and PUB (Singapore's national water agency) have brought together some of the world's leading experts to discuss this complex topic from multidisciplinary, multi-sectoral and multi-issue perspectives. This special issue of

PREFACE

the *International Journal of Water Resources Development* is a direct result of this meeting. I am confident that this issue will indeed contribute to finding implementable and socially acceptable solutions to a complex and difficult problem.

Asit K. Biswas
Lee Kuan Yew School of Public Policy, National University of Singapore

Introduction

In an increasingly globalized world, societies have become less resilient in regard to their natural environment. Long-term changes such as economic development, population growth and urbanization, as well as the impending threat of climate change, have increasingly resulted in global impacts on natural resources. Water resources, in particular, have become more polluted, mismanaged, misgoverned and poorly allocated.

As water demand has increased for numerous uses and users, and resources have become more limited because of physical scarcity, over-exploitation and pollution, it is therefore necessary to develop more options for water supplies. These options include the production at large scale of so-called non-conventional sources of water, such as recycled water from municipal sources, and desalination of seawater and brackish groundwater.

Appropriate planning and management consideration and improved treatment technologies can result in the production of high-quality water with no negative health or environmental impacts. Their economic, social and environmental benefits are many, as they address water supply scarcity, efficient resource use and environmental and public health considerations. Overall, they are used for potable and non-potable purposes, either directly or indirectly. Their usages include agriculture, landscape, stream and groundwater augmentation and managed aquifer recharge, cooling water for power plants and oil refineries, processing water for mills, toilet flushing, dust control, construction activities, concrete mixing and artificial lakes (United States Environmental Protection Agency [USEPA], National Risk Management Research Laboratory & U.S. Agency for International Development, 2012).

According to the US Environment Protection Agency, direct potable reuse is

> the introduction of reclaimed water (with or without retention in an engineered storage buffer) directly into a drinking water treatment plant, either collocated or remote from the advanced wastewater treatment system. Indirect potable reuse (IPR) is the augmentation of a drinking water source (surface or groundwater) with reclaimed water followed by an environmental buffer that precedes drinking water treatment. (USEPA, National Risk Management Research Laboratory & U.S. Agency for International Development, 2012, pp. 1–2)

However, terminology can be different in other countries. The terms 'reused', 'recycled' and 'reclaimed' water are also not always used interchangeably. For example, in the United States, reused water is known in different states as 'recycled' or 'reclaimed' water (Miller, 2006); in Singapore, it is known as NEWater (Lee & Tan, 2016).

Interest in the potential of reused water has increased globally, with studies focusing on policy, management, technology and public acceptance. With the best of the knowledge and experience available, governments, water utilities, academic institutions, non-governmental organizations and members of society are trying to develop or contribute towards development of guidelines and risks analyses; better understanding of economic aspects; safety, health and water quality considerations; social perceptions; environmental impacts and benefits; and more advanced and cost-effective technologies (e.g. ATSE, 2013; Eslamian, 2016; National Research Council, 2012). The analyses concur on the enormous potential of reused wastewater as a reliable source of clean water that also enhances urban resilience.

Rather than technological aspects, the main constraints on implementation of water reuse projects in different parts of the world seem to have been lack of full public support due to perceived health hazards and environmental impacts. A main handicap has been that governments and water utilities are usually slow to understand public concerns and perceptions. After backlashes in several places, public information, communication and awareness campaigns, broader participation and educational programmes have become integral parts of development policy and decision-making frameworks.

With the objective of better understanding water reuse, the Institute of Water Policy of the Lee Kuan Yew School of Public Policy, National University of Singapore, the National University of Singapore Environmental Research Institute and PUB, the Singapore national water agency, organized a workshop on water reuse policies for direct and indirect potable use, their associated global trends, health and environmental considerations and social perceptions and acceptance. The workshop was held at WaterHub, Singapore, on 15 and 16 June 2015.

Forty national and international experts participated in the workshop. Contributions were presented on how the water portfolios of different cities and countries have been diversified with high-quality reused water. These included Australia, China in general and Beijing in particular, Saudi Arabia, Singapore, the European Union, the United States and Windhoek in Southern Africa. Topics discussed included policies, legal and regulatory frameworks, guidelines, standards, water pricing, business models, participation of the private sector, health considerations, technological development and public perceptions, participation and acceptance. Also elaborated at the workshop were the potential of production failure and reliability and how to address them; how to communicate to the public; whether the global trends are expected to change over the short, medium, or long term due to scientific and technological developments; evolving global attitudes and social perceptions; and increasing water scarcity.

Robust papers based on the presentations and discussions during the workshop are published in this special issue on Water Reuse Policies for Potable Use.

The history of potable water reuse and frameworks for decision making, policy development and implementation, and the reasons for their success or failure, are topics argued at length. Ong (2015) discusses safety, health issues, water quality considerations, concerns of potential presence of pathogens and inorganic and organic constituents in the reused water and their health implications, and the need for specific or international guidelines or standards for treatment or monitoring when municipal wastewater is used for potable purposes.

Two most relevant examples are presented: Windhoek (van Rensburg, 2016) and Singapore (Lee & Tan, 2016). Windhoek started producing high-quality effluents and distributing them for direct potable use 48 years ago, in 1968. Specific breakthroughs in wastewater technology (mainly membrane technology) have allowed reclamation of water for reuse. In addition, support at the political level, technical design of the treatment plants, maintenance programmes and the appropriate skill level of operating personnel, together with public acceptance, were noted to play important roles in supply of potable water to the population during the last almost five decades.

In the case of Singapore, water reclamation was introduced in 2003. It has been planned as a key element of the nation's water sustainability and self-sufficiency strategies. Notable in the case of the city-state has been its large-scale implementation and wide public acceptance. NEWater (as it is known locally) is part of an overall approach that has the objective to change the mindset of the population to consider reused water a long-term renewable source to significantly increase the available water resources.

Cases were cited where projects have failed not because of lack of appropriate know-how or technological development but because of lack of governmental support, fear of public perceptions, and lack of public support, e.g. in Queensland, Australia.

Horne (2015) summarizes important lessons learnt that can apply to every other case discussed. These include the need for approaches to urban water security that are forward-looking and risk-based; health and safety guidelines that are stricter, to ensure very high-quality effluents; and public information and management of community perceptions as integral parts of policy and decision-making frameworks to ensure long-term acceptance and viability.

With rapid changes in the human and natural environments, including scarce natural resources that have the potential to impose limits on socio-economic growth and affect livelihoods, non-conventional sources of water will have a more important role to play in the future. For this to be achieved, decision making, planning and implementation should not focus so much on urban resilience or water resilience, as that would limit water reuse. It is social resilience that should be at the centre of the development discourse. Only then will unceasing progress be achieved.

References

Australian Academy of Technological Sciences and Engineering (ATSE). (2013). *Drinking water through recycling*. Melbourne: Australian Water Recycling Centre of Excellence.

Eslamian, S. (ed.). (2016). *Urban water reuse handbook*. Boca Raton: CRC Press.

Horne, J. (2015). Policy issues confronting Australian urban water reuse. *International Journal of Water Resources Development*, 32 (4), 573–589. doi:http://dx.doi.org/10.1080/07900627.2015.1090901.

Lee, H. & Tan, T. P. (2016). Singapore's experience with reclaimed water: NEWater. *International Journal of Water Resources Development*, 32 (4), 611–621. doi:http://dx.doi.org/10.1080/07900627.2015.1120188.

Miller, G. W. (2006). Integrated concepts in water reuse: managing global water needs. *Desalination*, 187, 65–75.

National Research Council. (2012). Water reuse: potential for expanding the Nation's water supply through reuse of municipal wastewater, Committee on the Assessment of Water Reuse as an Approach to Meeting Future Water Supply Needs. Washington: The National Academies Press.

Ong, C. N. (2015). Water reuse, emerging contaminants and public health: state-of-the-art analysis. *International Journal of Water Resources Development*, 32 (4), 514–525. doi:http://dx.doi.org/10.1080/07900627.2015.1096765.

United States Environmental Protection Agency (USEPA), National Risk Management Research Laboratory and U.S. Agency for International Development. (2012). *2012 Guidelines for Water Reuse*. Washington: CDM Smith.

van Rensburg, P. (2016). Overcoming global water reuse barriers: The Windhoek experience. *International Journal of Water Resources Development*, 32 (4), 622–636. doi:http://dx.doi.org/10.1080/07900627.2015.1129319.

Cecilia Tortajada
Institute of Water Policy
Lee Kuan Yew School of Public Policy
National University of Singapore

Choon Nam Ong
National University of Singapore Environmental Research Institute (NERI)

Potable water reuse history and a new framework for decision making

Joseph A. Cotruvo

Joseph Cotruvo & Associates, LLC, Washington, USA

ABSTRACT
As populations and water demand increase, more sustainable water sources are needed. Wastewater reuse is a major opportunity. Treated wastewater is available for non-potable applications and drinking water production. Direct potable reuse and planned indirect potable reuse provide sustainable drinking water; other reuse applications can offset current drinking water uses at lower cost due to lower end-use quality requirements. There is some public reluctance to choose potable reuse, but planned reuse projects provide drinking water of higher quality than typical natural sources. Guidance is available to assure safe and high-quality reused water.

Introduction

Water has always been reused for non-potable (e.g., industrial cooling or irrigation) and often for potable applications. Potable water reuse is not a new concept, but it has become a prime opportunity to provide high-quality drinking water in water-short areas where alternatives are not sufficient for population needs (Asano & Cotruvo, 2004). With the advent and refinement of better water treatment technologies, it is capable of yielding more efficient use of available natural water while providing water that is at least as safe and probably safer than local surface waters that are treated by conventional drinking water technologies (National Research Council [NRC], 2012).

Types of potable reuse have progressed over the centuries. Indirect potable reuse (IPR) occurs when waste discharges enter upstream and the mixed streamwater and wastewater is transported downstream to a drinking water intake, where it is usually treated to drinking water standards by conventional treatment technologies. Direct potable reuse (DPR) does not have the intermediate environmental passage and the waste stream is treated to drinking water quality using advanced treatment technologies and put into distribution to consumers with or without an engineered storage buffer (Cotruvo, 2014). The progression of potable reuse over time is summarized as follows:

- Unplanned or deliberate IPR: Untreated or treated upstream surface water discharge downstream to a municipal drinking water plant.

- Planned IPR with groundwater recharge: Soil aquifer treatment or injection of highly treated water into recoverable aquifers.
- Planned IPR: Advanced treated wastewater with surface discharge to a water body or groundwater recharge.
- IPR/DPR: Advance treated wastewater is discharged to the entry point of a drinking water treatment plant, or post-treatment blending, or storage in surface or groundwater prior to distribution.
- Pipe-to-pipe DPR: Treated wastewater is discharged to drinking water distribution without an environmental but possibly with an engineered storage buffer.

Before the 1970s many surface water sources in the United States were seriously contaminated because of uncontrolled wastewater and industrial discharges, resulting in microbial and chemical contamination of many surface waters. Major Clean Water Act legislation to control waste discharges was being implemented, beginning in 1972. This included: universal secondary treatment requirements for municipal wastewater discharges to surface waters; listings of priority industrial pollutants; initial implementation of pretreatment regulations; and National Pollution Elimination Discharge permits. These treatment requirements have prohibited discharge of billions of pounds of pollutants annually into US surface waters (Cotruvo, 2014).

The federal Safe Drinking Water Act of December 1974 (US Environmental Protection Agency [USEPA], 2015) was being implemented, beginning in 1976. Regulations for traditional inorganic, pesticide, radionuclide and microbial contaminants were being written and implemented. Regulations for trihalomethanes, disinfection by-products of chlorination, were in effect, and regulations on volatile industrial chemicals were being developed. Regulations for managing underground injection practices were being implemented to protect drinking water aquifers. Pesticide registration requirements were being developed, driven in part by the need to protect drinking water sources from contamination.

The US Environmental Protection Agency (EPA) began exploring potable reuse in 1980 with an expert workshop entitled Protocol Development: Criteria and Standards for Potable Reuse and Possible Alternatives (USEPA, 1980). Its purpose was to examine the state of the science of potable water reuse and to assess water quality, best available treatment technology, reliability, analytical chemistry, microbiology, toxicology, and human health issues. Since then, acceptance of DPR and the state of the scientific knowledge and technology have improved significantly.

Water technology

The conventional drinking water treatment for surface water sources at the time was coagulation, flocculation, sedimentation, sand filtration and chlorine disinfection. Powdered and granular activated carbon had applications for taste and odour control. Ozone and membranes were in their early stages. Microbial analyses consisted of total coliforms and *E. coli* or faecal coliforms, and heterotrophic plate counts. There was limited information on viruses and *Giardia*, and early indication of concerns for *Cryptosporidium*; polymerase chain reaction (PCR) was nonexistent. Analytical methods for trace organic chemicals were being developed and beginning to be applied in wastewater and drinking water facilities. Parts-per-billion measurements of organic chemicals were beginning to be applied to drinking water, but occurrence databases were very limited.

In 1980, the EPA initiated its interests in exploring direct potable reuse by organizing an expert workshop to review the issues and provide guidance on future directions. The workshop recommendations were:

- standards to define potable water regardless of source
- comprehensive characterization of source waters
- more toxicology study including concentrate studies
- better treatment technology with improved reliability and redundancy
- more implementation of groundwater recharge
- more non-potable reuse options.

Impediments included institutional, legal, and water rights issues and social and public acceptance of reuse (USEPA, 1980; Cotruvo, 2014). Thirty-five years later, much progress has been achieved.

- The state of science and technical understanding in the water industry has made major progress.
- Wastewater management and pretreatment have been instituted and the quality of wastewater has improved.
- Drinking water standards exist for almost 100 contaminants, plus broad treatment requirements for filtration and disinfection, as well as guidelines for hundreds of chemicals.
- Sophisticated instruments and analytical methods now allow quantitation at parts per trillion and lower levels, including more real-time on-line methods.
- Treatment technology now includes biological activated carbon, MF, UF, NF, and RO membranes, and advanced oxidation and ion-exchange resins.

Examples of advanced IPR and DPR potable reuse projects now operating or being developed and their treatment components are listed below (Cotruvo, 2014)

Windhoek, Namibia

- 1969 – $FeCl_3$, coagulation, dissolved air flotation, rapid sand, GAC, Cl_2 NaOH, blend (now considered non-potable; irrigation only)
- 1997 – PAC, pre-O_3, $FeCl_3$/polymer, coagulation, dissolved air flotation, $KMnO_4$, rapid sand, O_3, BAC, GAC, UF, Cl_2 NaOH, blend (about 30% recycled water)

NASA – International Space Station

- Urine distillate and air condensate recovery
- Multifiltration, vapour compression distillation, catalytic reactor, ion exchange, iodine

Big Spring, Texas

- Chlorinated secondary effluent, granular media → MF, RO, advanced oxidation, blend → flocculation, sedimentation, granular media, disinfection → distribution (issue: high salinity in natural source)

Singapore NEWater

- Secondary treated wastewater, UF, RO, UV

Orange County, California

- Groundwater replenishment and seawater intrusion barrier, 100 million gallons per day (mgd)
- Secondary effluent, NH_2Cl, MF, cartridge filter, RO (three-stage), H_2O_2/UV advanced oxidation, strip CO_2, lime stabilization

Domestic/commercial potable recycling

- Pure Cycle, Colorado, 1976–1982 – household wastewater, grinder, biodisk/cloth filter, MF, IX, UV → household storage tank
- Tangent Company in Ohio is developing and evaluating a single-building DPR system.

(Abbreviations in the preceding: GAC = granular activated carbon; PAC = powdered activated carbon; BAC = biological activated carbon; MF = microfiltration; IX = ion exchange; UF = ultrafiltration; NF = nanofiltration; RO = reverse osmosis; UV = ultraviolet light.)

Water reuse policy development

The past development of wastewater treatment philosophy has been to collect, process and dispose of wastewater so as to not injure the water environment. That is now being radically changed to considering wastewater as an asset to be exploited for all of the recoverable resources it contains, including the water itself (Cotruvo, 2012). From a public policy perspective DPR is a merging of the traditionally separate processes and organizations of wastewater treatment and drinking water provision. That creates a need to integrate the thought processes of the two disciplines and retrain many operators. DPR is intended to augment existing drinking water supplies that are already subject to drinking water regulations. Because of the intensive treatment, DPR product water is likely to be comparable or higher in quality than the existing water supply.

Recycled drinking water is drinking water, and should be subject to the same quality requirements as any other drinking water. Although DPR water is derived from an impaired water source, the technology is specifically designed to produce water of higher quality than many conventional drinking water systems, and with greater management oversight and more safeguards. Regulators will need to be educated on the benefits, safety and opportunities provided by DPR so that regulatory requirements will be appropriate to the new technology and its capabilities, while reflecting the reality that day-to-day operations require a higher level of diligence and qualifications.

Most countries do not currently have specifications that apply uniquely to direct potable reuse; Australia is an exception (Australia, 2006). A few states in the US have initiated guidelines and standards development. The EPA has published a compilation of water reuse activities and information (USEPA, 2012), and the International Life Sciences Institute has published guidelines for potable reuse in beverage processing (International Life Sciences Institute [ILSI], 2013).

There are two recent and current activities aimed at providing state-of-the-art background information and a set of principles to assist entities in developing standards for process decisions and standards. In the US a small expert group has produced a comprehensive Framework for Direct Potable Reuse (WateReuse Research Foundation [WRRF], 2015). It was initiated by the WateReuse Association's National Regulatory Committee and cosponsored by the WateReuse Association, American Water Works Association, Water Environment Foundation and National Water Reuse Institute. And the World Health Organization has recently begun a process to develop potable water reuse guidelines, which are expected to be available for review in 2017 (World Health Organization [WHO], 2015).

Framework for direct potable reuse

The WateReuse framework provides a policy structure for decision making for the introduction, design, management and regulatory oversight of planned DPR. DPR is intended to supplement and enhance the availability of adequate high-quality water supplies while assuring compliance with regulatory requirements. The focus of the framework document is: (1) to provide a discussion of what guidance would be required in each of the three key components that make up a direct potable reuse programme; and (2) to assist decision makers in understanding the role DPR projects can play in a community's overall water portfolio. The three components require an understanding and approach to address regulations, technology, and public involvement.

(1) Regulations
    a. Identifying potential public health risks and appropriate measures for their mitigation
    b. Defining the elements of a regulatory permitting process as applied to DPR
    c. Establishing operator training and certification requirements to assure continued performance of the more complex technologies that would be utilized

(2) Technology
    a. Understanding the treatment technologies and their performance capabilities for the production of advanced treated water that is protective of public health
    b. Demonstrating treatment performance through operational and treatment reliability, water quality (monitoring), operation and maintenance programmes, source control programmes, and residuals management
    c. Employing appropriate multiple treatment barriers, including management, treatment and operations
    d. Identifying issues associated with blending advanced treated water with other water sources

(3) Public outreach and participation in the decision process
    a. Need for public education and participation and outreach to the public and community stakeholders
    b. Importance of project knowledge and acceptance by the public and decision makers.

Structure of the DPR framework document

The purpose of the framework is to define DPR approaches and to provide a methodology for assessing the topics and issues that need to be addressed in the development of DPR guidelines on providing safe drinking water (WRRF, 2015). This includes analysis and discussion of what is considered direct potable reuse; key components of a successful/sustainable DPR programme; public health and regulatory aspects; source control programmes; principles of wastewater treatment; advanced water treatment; management of advanced treated water; process monitoring; residuals management; facility operations; public outreach; and future developments. The principal issues are:

- What are the goals?
- What are the results of health assessments and investigations?
- What treatment-related log reduction values are needed to assure microbial safety?
- What would be the applicable water quality and treatment requirements?
- What requirements would be applicable to sewage as a third water source in addition to surface water and groundwater?

The goals would be to meet or exceed all applicable drinking water regulations and guidelines and even to go further and address detected contaminants without existing regulations and guidelines to assure overall safety of the product water.

The primary objective of water treatment is to provide water that is safe for all intended uses. 'Safety' is interpreted in several ways. Some believe that it means zero risk of adverse outcomes, which is impossible for any human situation, since all activities have some risk. In general, the term is intended to mean that no adverse outcomes are likely under the conditions of normal use. The latter is certainly the goal of any conventional water supplier, and DPR is actually designed to have less risk than most conventional water supplies because its approach is designed to provide more substantial and redundant treatment technologies. It is expected to be managed with greater specificity than most conventional supplies, because it relies on source water that is more contaminated than most source waters.

Epidemiological studies

Epidemiological studies focus on measurable comparisons of changes in disease incidence in the exposed population versus populations with lesser or zero exposure. Ecological epidemiology studies compare populations based upon aggregated data, and are used primarily for generating hypotheses that might need to be tested in more detailed studies. Analytic studies compare cohorts or compare cases and controls and have some potential for establishing causality if they are sufficiently robust and consistent with other information, including biological plausibility (WRRF, 2015).

However, the limited power of epidemiology studies means that they are not generally able to detect low incremental risks and differentiate them from background. Environmental exposures to chemicals from food, water and ambient sources are difficult to define, and care must be taken to control for variables that may confound the outcome or result in exposure misclassifications. Numerous factors including ethnic distribution, genetics, life style, economics and others can contribute confounders.

The principal health issue in both DPR water and conventional drinking water is acute microbial risks. In addition there are concerns about longer-term potential exposure risks that might be associated with disinfection by-products and other trace contaminants. Apart from detection of a small number of acute water-borne disease outbreaks in conventional drinking water supplies in developed countries with strong management programmes, most studies of other concerns have dealt with disinfection by-products, with limited definitive outcomes because of the apparent low risks that may exist and the many confounders that are encountered. Cancer risk studies would require 20–40 years of chronic exposure. Conducting epidemiology studies in DPR water supplies is likely to be even more difficult and less definitive and very likely to find no detectable effects, except for occasional false positives due to weaknesses in study design not properly addressing the confounders that are inherent in studies of low-probability events.

Ecological studies are not appropriate, and case control studies that might be capable of examining very low risks are very difficult, very expensive and unlikely to be productive. Some studies have been conducted of soil aquifer treated IPR without adverse findings. Numerous in vitro and in vivo studies of chemical concentrates from water have not produced adverse findings. It is impossible to prove a negative, so studies of trace chemicals in drinking water that do not validly find an adverse effect can, at best, provide estimates of upper-bound risks. That could be a useful project validation in some instances, although the risk of obtaining false positives due to data deficiencies and poor study design is a significant concern that could mistakenly seriously impair public acceptance of the water.

Managing microbial pathogens

The obvious dominant risk from recycling wastewater is the presence of pathogens that must be removed to protect drinking water consumers (WRRF, 2015). The several filtration and disinfection technologies that are considered for DPR are highly effective at removing bacterial, viral and protozoan pathogens, such that none would be expected to be detectable in treated water, and indicator organisms like total coliforms and *E. coli* and faecal coliforms would also be below detection. A number of log removal performance requirements have been proposed, and they all typically would achieve common benchmark risks of less than one waterborne disease case per 10,000 consumers per year (Regli, Rose, Haas, & Gerba, 1991) or disability adjusted life years of less than 10^{-6} per lifetime (World Health Organization [WHO], 2011a). They differ by the amount of additional margin of safety that might be demanded to prevent adverse effects in the event of a temporary treatment excursion (e.g., shorter contact time or dosage of a disinfectant, or a membrane failure). These low-probability events are addressed by multiple barriers designed into the treatment train that would prevent a catastrophic failure that would affect final water quality. So, it is important to judiciously choose performance requirements that achieve safe water but without excessive conservatism that just increases costs and produces more complex treatment trains that are less amenable to practical day-to-day operation.

Some examples of target log reductions

DPR treatment configurations are being considered that would conservatively achieve more than 20 logs of theoretical removal of pathogens. It is instructive to put that into proper context.

Table 1. Microbial reduction criteria for evaluation of treatment trains and protection of public health (from WRRF (WateReuse Research Foundation), 2013; Framework, 2015).

Microbial group	Criterion, $\log_{10}$ reduction	Possible surrogates	Source used to develop criteria
Enteric virus	12	MS2 bacteriophage	SWTR (US EPA, 1989a); CDPH (2011); NRC (National Research Council) (2012); NRMMC–EPHC–NHMRC (2008)
Cryptosporidium spp.	10	Latex microspheres, AC fine dust, inactivated *Cryptosporidium* oocysts, aerobic spores	Interim ESWTR (US EPA, 1998); LT2 ESWTR (US EPA, 2006); CDPH (2011); NRC (National Research Council) (2012); NRMMC–EPHC–NHMRC (2008)
Total coliform bacteria	9	Enterococcus	Conventional drinking water experience with conservative assumptions

As a starting point, the 2006 LT2 Enhanced Surface Water Treatment Rules (Federal Register 71:3, 5 January, 2006, www.federalregister.gov) dealing with control of *Cryptosporidium* spp. specify log removals that depend upon the concentrations detected in 24 monthly samples of raw water. Sufficient treatment is required to achieve 2 logs of removal by disinfection with a protected source containing <0.01 oocysts/L, 3 logs (conventional surface drinking water treatment including filtration and disinfection) for a source with <0.075; 4 logs for 0.075–1, 5 logs for 1–3, and up to 5.5 logs for >3 oocysts/L.

California has adopted very conservative regulations for groundwater-recharge IPR projects that might be extended to DPR. These require 12-log enteric virus reduction, 10-log *Giardia* cyst reduction, and 10-log *Cryptosporidium* oocyst reduction beginning from raw wastewater.

Texas uses case-by-case considerations and begins with secondary effluent rather than raw sewage. For example, the Wichita Falls DPR project requires 9-log virus reduction, 8-log *Giardia* cyst reduction, and 5.5-log *Cryptosporidium* oocyst reduction based on an assessment of the quality of the secondary effluent and pertinent regulations.

In 2013, an independent advisory panel convened by the National Water Research Institute for WateReuse Research Foundation Project 11–02 (WRRF, 2013) suggested the log removals in Table 1 as conservative values that would achieve less than 1/10,000 annual risk. A log reduction criterion for *Giardia* was not retained because the reduction of *Cryptosporidium* oocysts would assure the same or greater reduction of the larger *Giardia cysts*. Total coliform bacteria was added as a criterion because of their presence in water at concentrations greater than enteric pathogens and *E. coli*, and are easily monitored. The panel concluded that a nine-$\log_{10}$ reduction of total coliform bacteria in the combined wastewater and drinking water treatment processes would provide a *de minimis* annual risk of infection from *Salmonella* spp. and other enteric bacterial pathogens.

Benchmarks for chemical contaminants

Some trace chemical contaminants, including pharmaceuticals, can be present in parts per billion or parts per trillion in conventional drinking water and IPR and DPR water, but they provoke hypothetical concern from possible long-term exposure (WHO, 2011; NRC, 2012; Cotruvo, Bull,

Crook, & Whittaker, 2012; Bull, Crook, Whittaker, & Cotruvo, 2011). The lower the concentration, the lower the potential for chronic exposure risk. Numerous chemicals have quantitative national regulations and international guidelines. Among the guidelines are the WHO Guidelines for Drinking Water Quality (WHO, 2011a), EPA Drinking Water Health Advisories (USEPA, 2012), EPA Pesticides Benchmarks (USEPA, 2013), and the Australian Reuse Guidelines (Australia, 2006).

In addition, for other chemicals of concern that might be detected, numerical values could be produced from standard risk assessment methodologies, 'margin of exposure' calculations for pharmaceuticals, and 'threshold of toxological concern' techniques as a structure activity approach for chemicals with minimal available toxicological information.

Monitoring

Speed in producing data on treatment train process performance and water quality is an important element in DPR, and also valuable for other water supplies. Real-time monitoring for critical control point surrogates using online systems should be utilized as the technology permits to assure that the performance of the treatment train is closely scrutinized and adjustments are made rapidly as needed to assure the safety of the finished water. The expectation is that treatment processes would be selected based upon determining relationships between benchmark trace chemicals for that locality and chemical and other measurement surrogates that would be used for operational critical control point monitoring to demonstrate that the processes are functioning as designed.

US regulations

There are numerous regulations in the US that address drinking water and wastewater controls and that provide substantial coverage that could readily be extended to DPR produced from wastewater as a third water source. The Safe Drinking Water Act gives primary maximum contaminant levels and secondary maximum contaminant levels, disinfection, filtration and turbidity requirements, total organic carbon and filter backwash specifications and source protection directives (USEPA, 2015). The Clean Water Act gives source protection requirements for wastewater discharge that include industrial pretreatment, industrial effluent guidelines, National Pollution Elimination Discharge System permits, and ambient water quality criteria. These can be augmented by local requirements to assure that the DPR system will not be challenged by chemicals that would be refractory to the treatment train being employed.

Conclusions

The technology exists to provide safe drinking water through IPR and DPR, and numerous technological options exist. Microbial safety is paramount to protect against acute risks. Trace chemical contaminants are a hypothetical long-term exposure risk scenario. Neither of those types of contaminants is unique to recycled water. A high level of system reliability and multiple barriers is essential to assure safety. Wastewater discharge controls and pretreatment are essential to assure that the advanced water treatment process is not challenged by contaminants for which it was not designed.

DPR is frequently complex and more expensive, so where available water is not capable of meeting demands into the future all options should be evaluated, including new water sources, desalination of seawater or brackish groundwater, substitution of lesser-quality water for lower-level applications, and conservation. If a DPR approach is selected, a DPR project and conventional water systems have much in common, since they both produce drinking water, and uniform quality standards should apply to both, rather than standards that are unique to DPR. Microbial protection is the most important aspect to be managed for both DPR and conventionally treated drinking water supplies, because the risks are acute and potentially significant. As with conventional drinking water regulations, virtually all chemical contaminants that are of concern from long-term exposures would be managed, so detections of excursions of trace levels do not usually represent immediate risks and can be managed by adjustments without major disruption of the DPR system.

Regulations and benchmarks are available for hundreds of potential contaminants, and others can be produced as needed using standard risk-assessment principles. Numerous configurations of treatment technologies are capable of converting wastewater into safe drinking water, and many choices are available for application. Piloting is an essential step in the selection of appropriate treatments as well as an important means for training operators to manage the treatment processes. Source protection and multiple barriers are important elements in any drinking water project. Real-time on-line monitoring of critical control points for indicator surrogates is key for assuring that the process is functioning as required at all times, and allowing adjustments to be rapidly applied when needed.

Consumer acceptance is an important issue, but it is primarily dealt with when the public agree that additional water is needed, and that IPR or DPR is safe and acceptable, and are willing to finance the project. That results in the design and initiation of the treatment system. Confidence in the competence and credibility of the water supplier must be maintained. This is best developed by establishing a credible and direct relationship involving capability and honesty between the supplier and the public. In general, epidemiology studies, if attempted, should not be expected to provide information other than not exceeded upper-bound values, within the weaknesses of epidemiological applications for low-risk outcomes. They must be very carefully designed because they also always have the potential of producing false positives that can create disruptions of public confidence.

Less than 1% of publicly supplied tap water is actually consumed as drinking water and cooking water where high quality is essential, but the public currently exercises more drinking water options by choice. Thus, it not critical whether a high percentage of the public actually drinks the water. That issue exists in conventional water supplies in developed countries, where many people choose to use beverage water such as bottled or point-of-use treated water in the home, so it should not be a determinant of success. Other papers in this volume address the question of public acceptance of potable recycled water and the significance of education, perception and emotion in the choice of what to drink.

Disclosure statement

No potential conflict of interest was reported by the author.

References

Asano, T., & Cotruvo, J. A. (2004). Groundwater recharge with reclaimed municipal wastewater: Health and regulatory considerations. *Water Research, 38*, 1941–1951.

Australia. (2006). *National water quality management strategy australian guidelines for water recycling: managing health and environmental risks*. Retrieved from http://www.australianwaterrecycling.com.au/guidelines

Bull, R. J., Crook, J., Whittaker, M., & Cotruvo, J. A. (2011). Therapeutic dose as the point of departure in assessing the health hazards from drugs in drinking water and recycled municipal wastewater. *Regulatory Toxicology and Pharmacology, 60,* 1–19.

Cotruvo, J. A. (2012). *Beneficial wastewater reuse: An idea whose time has come.* Water/Waste Processing, February, 6–9 [Editorial].

Cotruvo, J. A. (2014). Direct Potable Reuse: Then and Now. *World Water: Reuse & Desalination, 5,* 10–13.

Cotruvo, J. A., Bull, R. J., Crook, J., & Whittaker, M. (2012). *Health Effects Concerns of Water Reuse with Research Recommendations*. WRF 06-004. Alexandria, VA: WateReuse Foundation.

International Life Sciences Institute. (2013). *Water recovery and reuse: guideline for safe application of water conservation methods in beverage production and food processing*. Retrieved from http://www.ilsi.org/ResearchFoundation/RSIA/Pages/GuidelinesforWaterUse.aspx

National Research Council. (2012). *Water reuse: Potential for expanding the nation's water supply through reuse of municipal wastewater*. Washington, DC: National Academies Press.

Regli, S., Rose, J. B., Haas, C. N., & Gerba, C. P. (1991). Modeling the risk from giardia and viruses in drinking water. *J. American Water Works Association, 83,* 76–84.

USEPA (US Environmental Protection Agency). (1980). *Protocol development: Criteria and standards for potable reuse and feasible alternatives*. Presented at the workshop of USEPA, Warrenton, VA, July 29–31. Eds. EPA 570/9-82-005.

USEPA. (2006, January). *LT2 enhanced surface water treatment rule*. FR 71:3.

USEPA. (2012, September). *Guidelines for water reuse*. EPA/600/R-12/618. Retrieved from http://nepis.epa.gov/Adobe/PDF/P100FS7 K.pdf

USEPA. (2013). Benchmarks. Retrieved 9 October 2015 from www.epa.gov/oppfead1/cb/csb_page/updates/2013/hh-benchmarks.html

USEPA. (2015). *Safe drinking water act (sdwa)*. Laws and Regulations. Retrieved from http://water.epa.gov/lawsregs/rulesregs/sdwa/index.cfm

World Health Organization. (2011a). *Guidelines for drinking water quality*. Fourth Edition. Geneva.

World Health Organization. (2011b). *Pharmaceuticals in drinking water*. Retrieved from http://www.who.int/water_sanitation_health/publications/2011/pharmaceuticals/en/

World Health Organization. (2015). Water sanitation health. Retrieved from www.who.int/water_sanitation_health

WateReuse Research Foundation. (2013). *Independent advisory panel final report: Investigating the criteria for direct potable reuse*. Publication Number NWRI-2013-01. Fountain Valley, California: National Water Research Foundation.

WateReuse Research Foundation. (2015). *Framework for direct potable reuse*. Independent Advisory Panel Report. WateReuse Project Number: 14–20. Alexandria, VA: WateReuse Research Foundation.

Water reuse, emerging contaminants and public health: state-of-the-art analysis

Choon Nam Ong

NUS Environmental Research Institute (NERI) & Saw Swee Hock School of Public Health, National University of Singapore, Singapore

ABSTRACT

This article addresses the issue of quality in reused water for potable purpose. The concerns of potential presence of pathogens and inorganic and organic constituents in the reused water and their health implications are discussed. Presently, there are no specific or international guidelines or standards for treatment or monitoring when municipal wastewater is used for potable purpose. Research to advance the safety, reliability and economic sustainability of reuse is also lacking. When assessing the risks associated with reclaimed water, the potential of production failure and reliability also needs to be addressed and mitigated.

Introduction

Freshwater is a precious resource. The ocean holds 97% of the earth's water; the remaining 3% is freshwater that is mostly inaccessible. Of the 3% of water not found in the ocean, about 69% is frozen in glaciers and icecaps. Only about 1% of all the freshwater on Earth is available to sustain and support life on our planet. With the world's population expected to rise from 7 billion in 2014 to 9 billion by 2050 and continuous industrial growth across the globe, water's essential nature has made it a strategic natural resource globally. Currently, over 700 million people worldwide lack access to clean water (WHO/UNICEF, 2015). In view of the multiplicity of factors that pose serious challenges to fresh water supplies, such as climate change, depleted groundwater and pollution of water sources, the problem of water shortages has become a matter of serious concern that needs urgently to be addressed. The development of appropriate methods to supply stable, sufficient and safe drinking water has become imperative.

Over the years, there has been a significant increase of water reuse in various parts of the world as an alternative source of drinking water. Water reuse is the practice of using water that has already been used. The terms 'reclaimed water', 'recycled water' and 'reused water' are often used interchangeably. Reused water can be defined as wastewater treated or processed to a certain standard suitable for reuse. This recovered water offers a new source of water that can be used for many different applications. However, it is important to note that the quality must be tailored to the requirements of the end use (Ong, 2015). The main objective of this article is to briefly review the current technologies, water quality and potential health issues related to potable reuse.

Types of water reuse for potable purpose

From an environmental point of view, water recovery for reuse is a unique approach in which water can be recovered and treated to meet certain quality level for use in the same or other applications. This conservation-*cum*-recycling approach can reduce total water consumption and result in less waste, which in turn fosters improved sustainability as well as continued offerings of a high-quality asset.

There are two major types of water reuse for drinking: direct and indirect. Direct reuse refers to the introduction of recycled water from a water reclamation plant into a water distribution system, i.e. treating used water and then piping it to water works that channel water directly to a housing estate for potable use. In indirect reuse, the treated water is added to a water supply source such as a reservoir, lake, river or aquifer, allowing it to be used again.

Historically, unplanned indirect reuse of wastewater in the water supply has been common in countries with long riverine systems. Some drinking water treatment plants have used water of which a large fraction originated as wastewater effluent from upstream communities on the same river. Cities along other major rivers have used water that was used by residents upstream, put back into the river, and then reused, in most cases repeatedly (Ong, 2015). For example, the city of London receives 20% of its drinking water from a tributary of the Thames. This is also the case for many other cities along the Rhine and Danube Rivers in Europe, the Ohio, Colorado and Mississippi in the US, and the Yangtze and Huangpu in China. For many years, these cities have inadvertently been practising indirect potable reuse. As such, the quality of the water is much more important than the type of the water source.

Today, with advanced treatment technologies becoming increasingly available, water can be treated to stringent quality guidelines for the intended use. The reclaimed water can be used in numerous applications to satisfy most specific water uses, depending on the level of treatment. Presently, reclaimed water is more commonly used for non-drinking purposes, such as agriculture, landscape, horticulture and park irrigation. Other major applications include greywater for cooling towers, power plants and oil refineries, toilet flushing, dust control, construction activities, concrete mixing, and artificial lakes.

A planned reuse system involves treatment of wastewater to certain standards to be reused with predetermined purposes. There are several existing planned potable reuse projects in the world. Past research and operational data from indirect potable reuse (IPR) facilities indicate that available technology can reduce chemical and microbial contaminants to levels comparable to or lower than those present in many current drinking water supplies. Discharging treated wastewater effluent to a natural environmental buffer, such as a stream, reservoir or aquifer and subsequently using it for drinking has been deemed an appropriate practice.

A number of these planned IPR projects have been ongoing for many years, demonstrating successful operations management and the observance of good drinking water quality guidelines. The world's first potable use facility has provided drinking water in Windhoek, Namibia, since 1968. In Virginia, the Upper Occoquan Service Authority launched one of the US's earliest indirect potable schemes in 1978, returning reclaimed water to a reservoir.

Today, water reuse has proven to be a sound option and shown to be cost-effective in many different parts of the world.

In contrast to IPR, direct potable reuse (DPR) is planned or intentional usage of treated wastewater as a direct drinking water supply. In some parts of the world, DPR may be a more economical and reliable method of meeting water supply needs than IPR. While DPR is still an emerging practice, it should be evaluated carefully in terms of water management planning, particularly for alternative solutions to meet water supply requirements that could be energy-intensive and ecologically unfavourable.

Notwithstanding the demonstrated safety of using highly-treated reclaimed water for IPR or DPR, there is still a need for more research to further advance the safety, reliability and cost-effectiveness of potable reuse as it relates to public health protection.

Quality of water: health considerations for potable purpose

After human influence, wastewater often contains a wide range of inorganic, organic and biological constituents. These constituents may impact human health and/or ecosystems, contingent on the concentration, duration and path of exposure (NRC, 2012). Some essential micronutrients, such as trace elements usually present at lower concentrations, may become toxic at higher concentrations. A constituent may have an adverse impact on aquatic species but no adverse impact on human health. Water reuse involves multiple potential applications, and the constituents of concern depend upon the final use of the water. For example, some constituents in reused water that may affect human health may not be of concern if the water is used for agriculture, irrigation or industrial applications where risk to human health from incidental consumption is negligible. This article focuses on domestic wastewater that is reused or treated and discharged to a water body that will later be a source of municipal water supply and used for human consumption. It is also important to remember that the occurrence and concentration of these chemicals and microorganisms are likely to vary from one source to another as well as from one location to another, with the treatment methods applied, and according to post-reclamation storage and conveyance practice. Depending on the reuse application, influent wastewater will need to be routed through differing degrees of water reuse system designs to remove the targeted impurities. As mentioned in the next section, a wide diversity of potential pathogens and chemical contaminants can be present in reclaimed water, and it is thus important to keep in mind that there are often other sources which could also contaminate the source water. To ensure the quality of reclaimed water for potable purpose, treatment systems should include multiple barriers for pathogens that cause water-borne diseases to strengthen the reliability of contaminant removal, and should employ diverse combinations of technologies to address a broad variety of contaminants (Figure 1). Presently, the four core stages of wastewater treatment are preliminary, primary, secondary, and tertiary (or advanced) treatment. The number and types of treatment can be selectively employed depending on how the water will eventually be used. Most recycled water, however, will undergo some form of disinfection (Figure 1).

Traditionally, chlorine has been used as a disinfectant. However, ozone and UV radiation have rapidly become alternative methods of disinfection (Table 1). Ozone and UV radiation are both more effective disinfectants than chlorine, but both are energy-intensive and costly. To achieve adequate disinfection, a combination of UV with additional disinfection agents can be used to provide higher reliability and efficiency in inactivation of different types

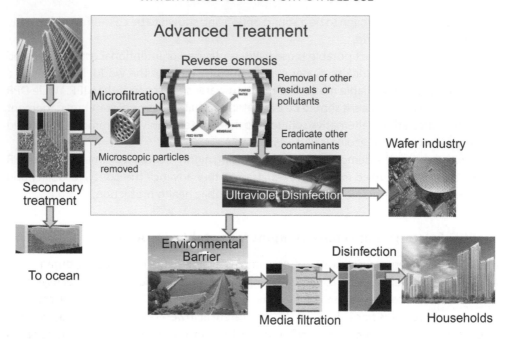

Figure 1. Reuse of water for potable purpose. Treatment schemes vary, but all require multiple barriers.

of microorganisms. Ozone is as effective as UV and chlorine, but like chlorine may lead to formation of secondary disinfection by-products (Jeong et al., 2015; Pals, Ang, Wagner, & Plewa, 2011; Plewa et al., 2012).

Despite the use of advanced technologies or treatment methods, reclamation facilities should develop monitoring and operational plans to respond to variability, equipment malfunction and operator error to ensure that the reclaimed water meets the appropriate quality standards for potable purpose. A critical aspect of such systems is the identification of measurable parameters, either chemical or bacteria indicators, as surrogate markers that can indicate treatment failures and trigger corrective actions. Some of the factors that can affect the quality of reclaimed water for potable purpose are described next.

Table 1. Log reduction efficiency of microorganisms from source water.

Treatment	E. coli (indicator bacteria)	Enteric bacteria	Enteric viruses	Giardia lamblia	Cryptosporidium parvum	Helminths	Phage (indicator virus)
Secondary treatment	1–3	1–3	0.5–2	0.5–1.5	0.5–1	0–2	0.5–2.5
Membrane filtration (UF/NF, RO)	4–>6	>6	2–>6	>6	4–>6	>6	2–>6
Ozonation	2–6	2–6	3–6	2–4	1–2	Data not available	2–6
UV disinfection	2–6	2–>6	1–>6	3–>6	3–>6	Data not available	3–>6
Chlorination	2–>6	2–>6	1–3	0.5–1.5	<0.5	<1	<2.5
Reservoir storage	1–5	1–5	1–4	3–4	<3.5	3	1–4

Notes: Reduction efficiencies depend first on source, treatment conditions and quantities of chemicals used, and second on pore size, types of filter, pretreatment process, efficiency and life-spent of membrane. They also depend on the installation, usage and maintenance of the membranes.

Pathogens: microorganisms

Wastewater from households contains a myriad of microorganisms, of which a fraction are potential human health hazards, in particular the enteric pathogens (NRC, 2012). Classes of disease-causing microbes are bacteria, viruses and parasites, such as helminths and protozoa. Craun, Craun, Calderon, and Beach (2006) report that in the United States, the enteric protozoa *Cryptosporidium* and *Giardia*, the enteric bacteria *Salmonella*, *Shigella* and toxigenic *Escherichia coli* O157:H7, and the enteric viruses, enteroviruses and norovirus, are the most commonly identified water-borne enteric pathogens. These pathogens can bring about gastrointestinal illness and have the potential to cause vast human mortality worldwide.

Bacteria are single-celled prokaryotes which are ubiquitous in the environment, including water. Domestic wastewater contains numerous bacteria, including pathogenic and beneficial types. Pathogenic bacteria cause gastrointestinal infection and are transmitted through the faecal–oral route. Examples of bacteria of faecal origin are the coliforms and enterococci species. Owing to their public health significance, monitoring systems and water quality standards have been established for faecal coliforms and enterococci in most parts of the world.

Viruses are tiny infectious agents, and they are of special concern in potable reuse applications because of their small size. Some of them are resistant to antibiotics, and thus are challenging for water reuse quality issues.

The recent National Research Report (NRC, 2012) provides a list of microbial agents associated with water-borne disease outbreaks and includes agents in wastewater thought to pose significant risk to humans. Table 1 briefly summarizes the efficiency of various methods of treatment in removing water-borne pathogens, including parasites, bacteria, *Giardia*, and *Cryptosporidium parvum*. As can be seen in this table, integration of advanced technologies including ultrafiltration and reverse osmosis appears to be the most effective in removing pathogenic microorganisms. Disinfection using physical and chemical agents such as UV, ozone and chlorine could further eliminate residual potential pathogens. Retention in an environmental buffer, such as an aquifer or reservoir, allows water renaturalization (where the water returns to its natural state in rivers and reservoirs) and provides an additional opportunity to attenuate microbial contaminants.

Organic chemicals

Organic matter is the major pollutant in wastewater. The vast majority is natural organic matter and soluble microbial products, mixed with small concentrations of a variety of individual organic chemicals. These trace organic chemicals originate from domestic or industrial activities and may contain pharmaceuticals and personal care products (PPCPs), surfactants, flame retardants, and also pharmaceuticals and their metabolites or residues, as well as steroidal hormones that are excreted by humans. Many of these organic constituents can be detected at low concentrations, generally micrograms per litre or less, during the water treatment process. Ram (1986) reports that over 2000 organic chemicals have been identified in nanogram per litre to microgram per litre concentrations in water around the world, including more than 700 in finished drinking water. As the technology for chemical analysis has advanced rapidly over the past two decades, the capability to identify and measure trace levels of organic chemicals in environmental media, particularly water, has also increased significantly. As a consequence, there are many reports of a wide range of substances, mostly anthropogenic contaminants, being identified in wastewater as well as in drinking water, usually at concentrations in the low nanogram per litre range. Numerous reports from different parts of the world also show pharmaceuticals, PPCPs and perfluorinated compounds (PFCs) of different levels in source water (Daughton & Ruhoy, 2009), river water (Kunacheva et al., 2011), and drinking water (Cooney, 2009; Kuster et al., 2008). The recent NRC report (2012) notes that a wide range of trace organic contaminants, both natural and synthetic, can be detected in reclaimed water. They include industrial chemicals such as PFCs, 1–4 dioxane, commonly used pesticides, natural and synthetic hormones, geosmin from algae, nanomaterials from sunscreen, and various types of fire retardants. Antibacterials and commonly used drugs such as beta blockers can also be detected.

As mentioned above, new analytical capabilities have facilitated the identification and quantification of new chemicals; this has led to the coining of the term 'emerging contaminants'. The US Geological Survey (USGS, 2012) has defined emerging contaminants as "any synthetic or naturally occurring chemical and/or microbial constituents that have not historically been considered to be contaminants". The presence in water of these substances, some of which are known to be biologically active, raises questions regarding their significance to aquatic organisms and, particularly, drinking water and what needs to be done to mitigate any significant risks to the environment and to human health. Many of these substances have probably been present for a long time, probably over many decades, and it is only now that they can be identified and quantified in water (EPA, 2012; Fawell & Ong, 2012). Often these groups of compounds are not covered by current regulations as the risk to human health, frequency of occurrence, source, etc. may not be known. As such, the term 'emerging contaminants' could be misleading and it was later adjusted to 'contaminants of emerging concern' (CECs). The US Environmental Protection Agency (EPA, 2015) defines CECs as "pollutants not currently included in routine monitoring programs" that "may be candidates for future regulation depending on their ecotoxicity and potential health effects and frequency of occurrence in environmental media". From the health risk perspective, the challenge is not so much their detection but rather the determination of their human and environmental health relevance.

The contribution of potential risk from drinking water can differ among compounds and depends on their concentrations relative to those in other matrices, exposure volumes, and efficiency of uptake after exposure. As mentioned above, modern analytical technology allows the detection of chemical and biological contaminants at extremely low concentrations, but the detection of a contaminant in reclaimed water does not, in and of itself, indicate a significant health risk. Although at this stage it is not possible to be absolutely sure that there are no effects resulting from human exposure to trace levels of CECs in drinking water, any effects would be extremely difficult to detect against the background of natural disease in the human population. In addition, a recent comparative study on exposure to estrogenic activity and trace contaminants in US municipal drinking water with food, beverage and air demonstrates that water consumption represents only a small fraction of exposure to pharmaceuticals, personal care products and endocrine disruptors (Stanford, Trenholm, Holady, Vanderford, & Snyder, 2010). Further, the study further no clear evidence of adverse human health effects based on the concentrations present in US drinking water and consumption patterns. Population studies done in the US and Europe also support the conclusion that the magnitude of risk for chemicals in drinking water tends to be small and there are no direct human health consequences (Bull, Crook, Whittaker, & Cotruvo, 2011; Mons, Hoogenboom, &

Noij, 2003). Several other groups have performed exposure assessments on the contribution of drinking water for the general population for PFCs, such as perfluorooctanoic acid (PFOA) and perfluorooctane sulfonate (PFOS) in Germany and for estrogenic hormones in the USA. It was shown that exposure was limited to 0.7–2% of the total daily intake of the general population (Fromme, Tittlemier, Völkel, Wilhelm, & Twardella, 2009). If we take pharmaceuticals as an example, they are normally governed by relatively stringent regulatory processes and require preclinical and clinical studies to assess their efficacy and safety to a greater extent than many commercial products. Thus they are generally better characterized than other environmental contaminants, such as PPCPs or some endocrine disruptors. Thus, reported concentrations of individual pharmaceutical compounds in drinking water, except perhaps in one or two specific locations of exceptionally high input and no treatment (Lubik, 2009), are too low to cause acute or even sub-acute effects (Bull et al., 2011; WHO, 2012). However, there are gaps in our knowledge with regard to very long-term exposure to substances normally used only for short-term treatment and to some population subgroups such as the foetus or developing child not normally exposed to some substances. With regard to CECs, the evidence is that concentrations in drinking water, if detectable, are normally too low to be of any concern (Fawell & Chipman, 2000; Wenzel, Muller, & Ternes, 2003; WHO, 2012). However, where there is only minimal sewage treatment and/or minimal water treatment the extent of exposure remains unclear (Fawell & Ong, 2012).

Nanomaterials in source and treated water

In recent years, nanomaterials synthesized via engineering methods, referred to as engineered nanomaterials (ENMs), are increasingly being utilized in the manufacture of common consumer and household products such as cosmetics, paints, hygienic coatings, textiles and building materials. ENMs are a class of small-scale substances generally defined by a particle size of less than 100 nm. ENMs include both nanoparticles and nanostructured materials (nano-sized topography or similar features). Their tiny size bestows these materials with properties different from those of larger particles. These properties include a large ratio of surface area to mass, resulting in increased reaction kinetics and special optical and electrical properties. The smallest engineered nanomaterials (nanoclusters) are smaller than 2 nm and exhibit excellent disinfection capability, and are thus attractive for many applications, ranging from cosmetics to medicines. The use of these materials has increased tremendously, but our understanding of their environmental and health effects has not been able to catch up with this rapid technological development. Recent data suggest that some highly commercialized ENMs will ultimately lead to appreciable concentrations in environmental compartments, including water, air, sediments and soils (Bernhardt et al., 2010). To maintain sustainable growth in this industry and ensure that public health is not at risk, there is now a worldwide research effort to better understand the environmental fate, bioavailability and exposure risks of ENMs. Presently, data on environmental occurrence and concentrations are extremely sparse. Whether these materials are present in source waters and treated water at concentrations that could cause concern, a call for detection of their existence in the water cycle to establish a database is needed. Given the diversity of nanomaterials, it is difficult to draw general conclusions regarding their toxicological effects. Further, the environmental fate and transport of nanomaterials in our water cycle depend on a number of factors, including aggregation, adsorption, abiotic degradation and biosystem-mediated changes. They could also serve as transport vehicles for other contaminants. One of the important aspects is that most common or even advanced drinking water treatment processes are not designed specifically to remove nanoparticles, although some removal would occur. Aggregation makes removal more likely. It has recently been shown that natural or synthetic organic matter may play a role in this process. So far, studies are lacking on the ability of membranes to remove nanoparticles, although reverse osmosis and nanofiltration would be expected to remove some nanoparticles and need to be further explored. Kaiser et al. (2009) investigated the removal of titanium particles at wastewater treatment plants and found that the majority of titanium in raw sewage associated with particles of size greater than 0.7 μm were generally well removed through a process train; whereas particles of less than 0.7 μm were found in the treated wastewater effluent. This finding suggests that particles of nano-size may not be effectively removed by membrane technology.

Currently, there are no guidelines or regulations in the use of nanomaterials, although it is acknowledged by many regulators that there is a need to have certain guidelines. It is also worth mentioning that decisions and standards have to based on sound exposure and health effects, for which there is now a significant knowledge gap. It is heartening to know that internationally, numerous countries are working on regulation of nanotechnologies, either on specific guidelines or relevant provisions; they include Australia, Canada, the European Union, Japan, the United Kingdom and the US EPA.

In brief, in developing the water infrastructure of the future, it is imperative that the potential risk to public health stemming from xenobiotics, including nanomaterials, is assessed. The knowledge obtained is critical for developing technologies to minimize their presence, for early prevention of risk and for guideline development. Information on the presence of ENMs after reverse osmosis and the bioavailability to aquatic biota is also currently lacking. In addition, methods for their early detection have so far not been fully investigated. Proper assessments of these concerns include identifying and quantifying their sources. Information on the concentration of a contaminant likely to cause health effects would allow scientists to determine whether the level of nanomaterials is of potential health risk.

Health considerations of long-term consumption of demineralized water

Regulatory guidelines that set out the maximum acceptable concentrations of inorganic and organic substances and microorganisms have been established internationally and in many countries to assure the safety of drinking water. However, these guidelines do not take into consideration the potential effects of totally demineralized water, since this water is not usually found in nature. At the present moment, demineralized water is not used in countries where drinking water regulations were developed; they are however used in some other communities. With respect to demineralized water, researchers are focused on two main issues (WHO, 2009, p. 1): the possible adverse impacts on human health; and the desired or optimum concentration of minerals in drinking water required to meet health considerations.

Demineralized water is devoid of minerals and as a result might not be fully suitable for human consumption. The water tends to be of lower pH, and if untreated its distribution through pipes and storage tanks would cause serious corrosion challenges. This aggressive behaviour may cause leaching of metals and other materials from distribution pipes and plumbing materials. Totally demineralized water such as distilled water has poor taste characteristics. There have been increased reports of the health value of minerals in water over the past several decades. Recent epidemiological studies report lower morbidity and mortality from certain cardiovascular diseases in areas with hard water, suggesting the importance of minerals to human health (Monarca, Kozisek, Craun, Donato, & Zerbini, 2009; Momeni, Gharedaghi, Amin, Poursafa, & Mansourian, 2014). The potential health effects of long-term consumption of demineralized water are of interest not only in countries lacking adequate fresh water but also in places where some types of home water treatment systems are widely used for water purification. Consequently, the exposures and risks should be considered not only at the community level but also at the individual or family level (Davies, 2015). Although people have been aware of the importance of minerals and other beneficial constituents in drinking water for thousands of years, it has received less attention in guidelines and regulations. International and national authorities responsible for drinking water quality may want to consider guidelines to specify the minimum content of relevant minerals and to promote research to establish these guidelines.

Natural systems for water treatment and the role of the environmental buffer

Currently, there are many treatment options available for water reuse for potable purpose, including combined multiple technologies and natural treatment processes. A multiple-barrier approach that can be tailored to meet specific water quality requirements appears to be most attractive for many reclaimed wastewater establishments. In most potable water reuse systems, water is discharged after treatment to a natural system such as an aquifer, a reservoir or a wetland, providing a buffer between water treatment and consumption. Environmental buffers can further remove contaminants such as pathogens from the water (Saxena et al., 2015) and provide additional retention time, and they also have been beneficial for public acceptance of water reuse. In some cases, engineered natural systems can replace advanced treatment processes. However, the science necessary to design engineered natural systems to provide a uniform level of public health protection is currently unavailable. Environmental buffers are useful elements that should be considered along with other processes and management actions in the design of potable reuse projects; but it is important to note that though they offer an additional barrier they do not necessarily provide quality assurance. Thus far, most of the advanced treatment technologies are capable of addressing pathogens or organic contaminants, with natural systems offering additional confidence and acceptance by the public.

Governance and quality assurance of water reuse for potable purpose

The availability of advanced multi-barrier treatment processes is indeed a major step forward in recycling water for potable purpose. However, to assure that the water produced meets the required standard, comprehensive water sampling has to be carried out on a regular basis, depending on the potential contaminants in both the source and treated water. Contaminants that are of significant health concern and found to be potentially present need to be assayed more frequently. Methods of detection will also need to be updated with the advancement of detection methods. Comprehensive physical, chemical and biological testing data, together with surrogate markers, are useful means for public communication for acceptance of water reuse. Utilities should also engage independent panels of experts to regularly audit and review plant performance, water quality variation and frequency of measurements. Emergency plans as well as alternate supplies should also be considered in the planning of water reuse programmes. A reliable and robust system will ensure the sustainability of a reuse project for drinking purpose.

Conclusion and the way forward

Water reuse with advanced technologies has been considered a viable, long-term solution to the challenges of population growth and industrial and agricultural demands for water and also to offer significant additional potential supplies to help meet future needs (NRC, 2012). Advancements in treatment technology have led many to believe that planned IPR will become more common. Moreover, the evidence so far suggests that the health risks of exposure to disease-causing microbes from wastewater reuse with multi-barrier technology are low. Indeed, in most cases, the risk of microbial contamination is significantly lower than in the existing water supplies. It is also heartening to note that efforts to address the potential risk of exposure to other potential contaminants have also been drawing attention, as water reuse becomes a major component of potable supplies. Technologies available so far have been shown to offer an acceptable alternative to augment our drinking water supply, though the current technologies still have certain shortcomings, such as concerns with respect to small biological and chemical molecules and compounds of emerging concern that can pass through filtration membranes, as mentioned above. On the other hand, newer methods and technologies are expected to continue to advance in the coming years to overcome some of these issues and concerns. Membrane-based filtration processes have been particularly attractive over the past two decades for potable reuse applications because of their ability to produce high-quality water. However, they are of relative high initial cost and limited life span, and subject to membrane fouling, which necessitates regular maintenance for optimal performance. Thus, sound governance, in combination with a robust and comprehensive monitoring system, is needed to ensure minimal risk for the public. Universally acceptable guidelines and standards that address water quality and health are desperately needed to foster a wider application of alternative water sources, which could ease water scarcity in many countries and megacities. The potential of reclaimed treated effluent is enormous. Although water reclamation and reuse are practiced in many countries, the current levels of reuse constitute a small fraction of the total volume of the municipal and industrial effluent generated. Water reuse provides a wide range of benefits for communities, which trans-lates into immense value for the public and the environment. An integrated concept which involves the convergence of diverse areas such as governance, health risks, regulation and public perception, presents a significant challenge to water reuse. Climate change is also expected to favour water recycling and reuse, as water recycling is widely recognized as a proven response to water scarcity which enhances sustainability by providing drought-proof alternative resources (Ong, 2015). Therefore, water reuse should be viewed as one of several alternative sources of new supplies of water or as an important part of any water resource management planning activity, especially in water-stressed regions.

In brief, while technology and water quality monitoring systems can promise the delivery of safe, sufficient and secure drinking water, gaining public acceptance through quality assurance is critical and can be a major hurdle.

Acknowledgements

The author wishes to thank the NUS Environmental Research Institute for the Secondment Fund and acknowledges the support of various agencies, including the Singapore National Research Foundation and the Environmental and Water Industry office. The author also appreciates the opportunities to collaborate with colleagues from the national water agency, PUB, over the past many years on water reuse. Thank you also to Dr Eileen Tan for editorial input on this manuscript.

Disclosure statement

No potential conflict of interest was reported by the author.

References

Bernhardt, E. S., Colman, B. P., Hochella, M. F., Jr, Cardinale, B. J., Nisbet, R. M., Richardson, C. J., & Yin, L. (2010). An ecological perspective on nanomaterial impacts in the environment. *Journal of Environment Quality, 39*, 1954–1965.

Bull, R. J., Crook, J., Whittaker, M., & Cotruvo, J. A. (2011). Therapeutic dose as the point of departure in assessing potential health hazards from drugs in drinking water and recycled municipal wastewater. *Regulatory Toxicology and Pharmacology, 60*, 1–19.

Cooney, C. M. (2009). Study detects trace levels of pharmaceuticals in US drinking water. *Environ Sci Technol., 43*, 551.

Craun, M., Craun, G., Calderon, R., & Beach, M. (2006). Waterborne outbreaks reported in the United States. *Journal of Water and Health*, 04, 19–30.

Daughton, C. G., & Ruhoy, I. S. (2009). Environmental footprint of pharmaceuticals: The significance of factors beyond direct excretion to sewers. *Environmental Toxicology and Chemistry, 28*, 2495–2521.

Davies, B. E. (2015). The UK geochemical environment and cardiovascular diseases: Magnesium in food and water. *Environmental Geochemistry and Health, 37*, 411–427.

EPA (2012). *Guidelines for water reuse*. Washington, DC: U.S. Agency for International Development. EPA/600/R-12/618.

EPA (2015). Water: Contaminants of emerging concern. http://water.epa.gov/scitech/cec/

Fawell, J. K., & Chipman, K. (2000). Endocrine disrupters, drinking water and public reassurance. *Water and Environmental Manager., 5*, 4–5.

Fawell, J., & Ong, C. N. (2012). Emerging contaminants and the implications for drinking water. *International Journal of Water Resources Development, 28*, 247–263.

Fromme, H., Tittlemier, S. A., Völkel, W., Wilhelm, M., & Twardella, D. (2009). Perfluorinated compounds – Exposure assessment for the general population in western countries. *International Journal of Hygiene and Environmental Health, 212*, 239–270.

Jeong, C. H., Postigo, C., Richardson, S. D., Simmons, J. E., Kimura, S. Y., Mariñas, B. J., … Plewa, M. J. (2015). Occurrence and comparative toxicity of haloacetaldehyde disinfection byproducts in drinking water. *Environmental Science & Technology*. doi: 10.1021/es506358x

Kaiser, M. A., Westerhoff, P., Benn, T., Wang, Y., Pérez-Rivera, J., & Hristovski, K. (2009). Titanium nanomaterial removal and release from wastewater treatment plants. *Environmental Science & Technology, 43*, 6757–6763.

Kunacheva, C., Tanaka, S., Fujii, S., Boontanon, S. K., Musirat, C., & Wongwattana, T. (2011). Determination of perfluorinated compounds (PFCs) in solid and liquid phase river water samples in Chao Phraya River, Thailand. *Water Science and Technol., 64*, 684–692.

Kuster, M., de Alda, M. J., Hernando, M. D., Petrovic, M., Martin-Alonso, J., & Barcelo, D. (2008). Analysis and occurrence of pharmaceuticals, estrogens, progestogens and polar pesticides in sewage treatment plant effluents, river water and drinking water in the Llobregat river basin (Barcelona, Spain). *Journal of Hydrology, 358*, 112–123.

Lubik, N. (2009). India's drug problem. *Nature, 457*, 640–641.

Momeni, M., Gharedaghi, Z., Amin, M. M., Poursafa, P., & Mansourian, M. (2014). Does water hardness have preventive effect on cardiovascular disease? *Int J Prev Med., 5*, 159–163.

Monarca, S., Kozisek, F., Craun, G., Donato, F., & Zerbini, I. (2009). Drinking water hardness and cardiovascular disease. *European Journal of Cardiovascular Prevention & Rehabilitation, 16*, 735–736.

Mons, M. N., Hoogenboom, A. C., & Noij, T. H. M. (2003). Pharmaceuticals and drinking water supply in the Netherlands. *Kiwa* (Report No. 2003.040 OCT 2003).

National Research Council. (2012). *Water reuse: Potential for expanding the nation's water supply through reuse of municipal wastewater*. Washington: The National Academy Press.

Ong, C. N. (2015). Water reuse. In J. Bartram (Ed.), *Routledge handbook of water and health*, chapter 39. Abingdon: Routledge.

Pals, J. A., Ang, J. K., Wagner, E. D., & Plewa, M. J. (2011). Biological mechanism for the toxicity of haloacetic acid drinking water disinfection byproducts. *Environmental Science & Technology, 45*, 5791–5797.

Plewa, M. J., Wagner, E. D., Metz, D. H., Kashinkunti, R., Jamriska, K. J., & Meyer, M. (2012). Differential toxicity of drinking water disinfected with combinations of ultraviolet radiation and chlorine. *Environmental Science & Technology, 46*, 7811–7817.

Ram, N. M. (1986). Environmental significance of trace organic contaminants in drinking water suppliers. In N. M. Ram, E. J. Calabrese, & R. F. Christman (Eds.), *Organic carcinogens in drinking water: Detection, treatment, and risk assessment* (pp. 3–31). New York, NY: John Wiley & Sons.

Saxena, G., Marzinelli, E. M., Naing, N. N., He, Z. L., Liang, Y. T., Tom, L., Mitra, S., … Swarup, S. (2015). Ecogenomics reveals metals and land-use pressures on microbial communities in the waterways of a megacity. *Environmental Science & Technology, 49*, 1462–1471.

Stanford, B. D., Trenholm, R. A., Holady, J. C., Vanderford, B. J., & Snyder, S. A. (2010). Estrogenic activity of us drinking waters: A relative exposure comparison. *J Am Water works Association, 110*, 55–65.

USGS (2012). Toxic substances hydrology program. Emerging contaminants in the environment. http://toxics.usgs.gov/regional/emc/index.html

Wenzel, A., Muller, J., & Ternes, T. (2003). Study on endocrine disruptors in drinking water. Final report to the European Commission. *ENV.D.1/ETU/2000/0083*.

WHO/UNICEF, Millennium Development Goals. 2015. Report on Water & Sanitation, Joint Monitoring Program, Geneva – "Water, sanitation and hygiene: WASH Post 2015" http://www.wssinfo.org/fileadmin/user_upload/resources/JMP-A5-English-2 pp.pdf

World Health Organization. (2009). *Calcium and magnesium in drinking –water; public health significance*. Geneva: World Health Organization.

World Health Organization. (2012). *Pharmaceuticals in drinking water*. Geneva: World Health Organization.

Public–private partnerships in the water reuse sector: a global assessment

David A. Lloyd Owen

Envisager Limited, Ceredigion, UK

ABSTRACT
Unlike contracts involving water provision, involving the private sector in water reuse projects is not seen as politically contentious. Water reuse remains a small element of public–private partnership contracts in general, but their frequency of use has increased, notably since 2005. These contracts are typically awarded in middle-to-high-income countries, and there is a relation between contract awards and water stress. In terms of population served, 5% of contracts were awarded in areas without water stress and 77% in areas with high water stress.

Introduction

Water is seen as an "uncooperative commodity" (Bakker, 2004), meaning that appropriate pricing (full cost recovery and sustainable cost recovery) and the involvement of the private sector in operating water services in any way are seen are seen by some as contentious (Bakker, 2010).

Such reservations do not appear to apply to the same degree in the cases of wastewater treatment and water reuse. With the exception of the Windhoek contract in Namibia (Lahnsteiner & Lempert, 2007), water reuse does not involve any direct contact with domestic customers. Indeed, by its nature, water reuse ought only to deliver positive outcomes for domestic customers, since it is concerned with securing water supplies and alleviating water scarcity. By providing additional water supplies for industrial and recreational customers, water reuse benefits water management, as directly usable potable resources can be reserved for domestic consumption.

The importance of these additional supplies is highlighted by the twin pressures of population growth and urbanization, along with the need to "achieve universal and equitable access to safe and affordable drinking water for all" by 2030 as mandated in 2015 by the United Nations (Sustainable Development Goal 6.1). The United Nations in its Sustainable Development Goals (Goal 6.1) has two impacts. Firstly, the degree of political risk is lower for water reuse projects than for water supply projects, and secondly, appropriate pricing is feasible, which in turn improves the ability of such projects to attract financing from both public and private sources.

This article aims to present a snapshot of the development of water reuse private-sector participation (PSP) up to the end of 2014 which can be used as a framework for future comparisons.

Review of the literature

The need for nontraditional water sources in areas facing water scarcity has been extensively reviewed. For example, Jimenez et al. (2008) survey the development of water reuse in its demand management context. The two principal challenges facing water reuse are public perception and financing. Hartley (2006) highlights the need to build and maintain trust in the service, which requires a high degree of public outreach and education. Chen et al. (2015) note that public acceptance remains appreciably higher for non-direct and non-potable applications than for direct potable use. A particular challenge for water reuse is to reconcile the need for tariffs that ensure these projects can be financed while encouraging the use of the reclaimed water (Molinos-Senante, Hernandez-Sancho, & Sala-Garrido, 2013). A number of instances have been noted where PPPs have played a significant role in reducing anticipated project costs and therefore assisting the reconciliation of the cost and tariff elements (Lazarova et al., 2013). Keremane and McKay (2009) note that for PPPs to work in water reuse schemes they need the appropriate regulatory framework to ensure water standards and public confidence in those standards, a pricing framework that makes it attractive for customers to take up nonconventional water sources, and the ability to demonstrate that the private-sector parties are acting in the customers' interests.

Indirect potable example: Singapore's NEWater

Singapore's Public Utilities Board (PUB) is owned and operated by the government of Singapore. While the water and sewerage systems and customer relationships remain wholly under PUB's management, PUB is open to working with the private sector in developing and operating new water and wastewater treatment assets where this is demonstrably "more efficient" (Khoo Teng Chye, personal communication, 2011).

PUB aims to fulfil up to 55% of Singapore's water demand from NEWater water reuse facilities by 2060. As of 2014, 30% of Singapore's water was provided by NEWater facilities at Bedok (built 2003), Kranji (2003), Ulu Pandan (2008) and Changi (2009). A second plant at Changi (Changi II) is anticipated to open in 2016 (PUB, 2014). Ulu Pandan, Changi and Changi II are operated by private-sector companies, under design-build-operate contracts. The recovered water can be used directly as non-potable water by industrial customers or indirectly for potable use by being delivered to a holding reservoir, where it can be extracted, treated and distributed in the water network. The non-potable use is significant as up to 70% of total water demand is expected to come from industrial customers by 2060 (PUB, 2014).

Direct potable example: Windhoek

All potable water resources within 500 km of the city of Windhoek in Namibia are fully exploited. Water reuse was adopted as a means of maximizing the use of these resources. The New Goreangab Water Reclamation Plant entered service in 2002 and by 2004 provided 26% of the city's water needs. The plant is unique in being the only direct potable water reuse facility in operation. The facility is operated as a PPP on a 20-year operations and management contract. No cases of water-borne disease have been attributed to the plant since it entered service, and with only 5% of the city's domestic customers using point-of-use water treatment this indicates general confidence in the quality of the water delivered by the facility (Lahnsteiner & Lempert, 2007). While a concession contract was originally considered, two separate contracts were used, one for construction of the facility and the subsequent operations and management contract.

Methods

Media Analytics Limited published *Private Sector Participation in Water* in 2015 (MAL, 2015), based on the company's private-sector contract databases. The author was given access to the databases as well as the report. The use of the company's Global Water Intelligence project categories, including 'water reuse', and contract notes allows identification of all contracts where water reuse is involved. Contract details are as of the start time of the facility contract. This means that subsequent moves towards water reuse in wastewater treatment PPPs are not included.

The database is driven by GWI's three trackers of projects under development, from first proposals to their formal award. These cover desalination, water reuse and PPPs. The water reuse tracker covers PPP and public-sector contracts, and for January 2015 the tracker noted 168 projects under consideration; 49 of these had been announced during the previous 12 months.

Using the UN Development Programme's classification, 34 projects were in high-income countries, 35 in upper-middle-income countries and three in lower-middle-income countries; 32 of the projects were in OECD-member countries, 28 in BRIC countries (Brazil, Russia, India and China), and 12 in other countries.

No PPP water reuse projects starting before 1995 were identified. Contracts were grouped into four five-yearly segments by start date: 1995–99, 2000–04, 2005–09 and 2010–14. Non-water-reuse contracts starting before 1995 were excluded.

The GWI database has details about 3310 PPP contracts, including contract renewals, 2714 being in the period between 1995 and 2014, including 975 for wastewater treatment plants and 248 for water treatment plants where treatment capacity data were available for new contracts as opposed to renewals. The contract database was examined for contracts directly categorized as water reuse contracts and wastewater treatment contracts that include an element of water reuse.

Contracts that did not include details about the facility's treatment capacity or population served were excluded. Six projects were excluded on this basis, five being water reuse projects and one a wastewater treatment plant.

Fifty-four water reuse projects and 18 wastewater treatment projects incorporating water reuse were identified, a total of 72 projects. The latter were identified among 975 wastewater treatment construction projects.

Contract coverage is given in terms of water recovery capacity (in m^3 per day) and people served. For people served, the populations are either provided directly in the contract database or have been adapted from the treatment capacity data and the per capita water consumption by the relevant urban population in the country. Population coverage may be direct (as in Windhoek) or indirect (as in Singapore), the later meaning that the reused water is provided for non-potable applications, and other sources are as a result made available for potable water applications.

Results

Contracts noted

Seventy-two contracts were identified, covering 29.44 million people, with a total water reuse capacity of 11.80 million m³ per day, or an average of 401 litres per capita per day. Outside the US and Australia, urban domestic water consumption typically ranges between 130 and 200 litres per capita per day, indicating that at least 50% of the reused water is used for non-domestic applications such as industrial process water and for amenity irrigation.

Development of water reuse PPP contracts over time

Table 1 outlines the development of water reuse PPP contracts over time. The first contract identified started in 1998. The 1995–2004 period can be seen as one when PPPs were starting to be used for water reuse. Contracts were awarded at a steady pace during 2005–14. In 2010–14, other areas such as desalination were affected by funding constraints in the wake of the global financial crisis (MAL, 2014), which does not appear to have affected the development of water reuse PPP contracts to the same extent. There is no material difference between project sizes in terms of people served or capacity over time.

Despite the narrow band of average contract capacities and populations served, the contracts identified contain a notably wide range of project capacities, from 52 m³ per day (a community water reuse project in the rural US) to 1,990,000 m³ per day (a wastewater treatment and reuse project serving Mexico City).

Economic development and water reuse PPP over time

Tables 2 and 3 show how water reuse PP has developed over time by country grouping and World Bank classification. When comparing the OECD and BRIC countries to the rest of the world, contract awards in the rest of the world took place later than in the OECD and BRIC groupings. This related to contract awards in the Gulf States and in Singapore, which in general have been later to develop PPP markets than the OECD and BRIC countries (Lloyd Owen, 2015).

It was evident that water reuse PPPs emerged earlier in high-income countries, with contracts emerging last in the lower-middle-income countries. This reflects the greater ability of higher-income countries to fund these projects, especially at the earlier stage of development, when both development and operating costs are typically higher.

In 2005–09, the upper-middle-income countries are prominent, driven by major projects in China and Peru. The comparatively high number of contracts in population terms in

Table 1. Water reuse public–private partnership contracts over time.

	1995–99	2000–04	2005–09	2010–14	1995–2014
Capacity (million m³ per day)	0.31	1.06	4.70	5.73	11.80
Population (millions)	0.74	3.20	13.31	12.19	29.44
Number	2	9	29	32	72
Average capacity	0.16	0.12	0.16	0.18	0.16
Average population	0.37	0.32	0.46	0.39	0.41

Table 2. Population served by water use PPPs over time, by country grouping (millions).

	1995–99	2000–04	2005–09	2010–14	1995–2014
OECD	0.74	2.66	3.15	6.21	12.76
Brazil, Russia, India and China	0.00	0.50	4.05	2.47	7.02
Rest of the world	0.00	0.06	6.12	3.50	9.67
Total	0.74	3.21	13.31	12.18	29.44

Table 3. Population served by water use PPPs over time, by World Bank classification (millions).

	1995–99	2000–04	2005–09	2010–14	1995–2014
High-income countries	0.74	2.39	4.05	5.98	13.15
Upper-middle-income countries	0.00	0.83	8.67	5.91	15.40
Lower-middle-income countries	0.00	0.00	0.60	0.29	0.89
Total	0.74	3.21	13.31	12.18	29.44

Table 4. Average contract population and capacity, by country grouping, 1995–2014.

	Average population	Average capacity
OECD	0.97	1.16
Brazil, Russia, India and China	0.61	0.46
Rest of the world	1.97	1.82
Total	1.00	1.00

Note: Average population and average capacity are calculated as a proportion of the global average.

Table 5. Average contract population and capacity, by World Bank category, 1995–2014.

	Average population	Average capacity
High-income countries	0.95	0.90
Upper-middle-income countries	1.08	1.13
Lower-middle-income countries	0.73	0.51
Total	1.00	1.00

Note: Average population and average capacity are calculated as a proportion of the global average.

2010–15 for high-income countries may reflect the delayed impact of the global financial crisis, making project financing harder to obtain in middle-income countries.

In Tables 4 and 5, the average size of a contract in population and capacity terms is examined as a proportion of the overall contract size. Contracts awarded in BRIC countries appear to be smaller than in other areas. This reflects the emphasis on medium-scale projects for industrial parks and economic development zones in China and India. No contract awards were noted in Russia or Brazil. The higher average size of contracts in the rest of the world reflects the effect of a disproportionately small number of major contracts in Singapore and Peru (just four in all, with a combined capacity of 2,090,000 m^3 per day).

There was no appreciable difference between contract sizes in high- and upper-middle-income countries. For lower-middle-income countries, the smaller contract size reflects the three contracts awarded in India, including small industrial/mixed-area contracts.

Tables 6 and 7 outline the average contract size over time by country grouping. While contract size in OECD and BRIC countries was relatively constant throughout the period, there appears to have been some variability in other countries. This is due to the effect of the small number of contracts involved and the effect of some substantial contracts starting in 2005–09.

Table 6. Average contract size in population over time, by country grouping (millions).

	1995–99	2000–04	2005–09	2010–15	1995–2014
OECD	0.37	0.38	0.31	0.48	0.40
Brazil, Russia, India and China	0.00	0.25	0.27	0.22	0.25
Rest of the world	0.00	0.06	1.53	0.50	0.81
Total	0.37	0.32	0.46	0.39	0.41

Table 7. Average contract size in capacity over time, by country grouping (million m^3 per day).

	1995–99	2000–04	2005–09	2010–14	1995–2014
OECD	0.16	0.14	0.13	0.27	0.19
Brazil, Russia, India and China	0.00	0.04	0.09	0.06	0.08
Rest of the world	0.00	0.02	0.51	0.22	0.30
Total	0.16	0.11	0.16	0.18	0.16

The main difference between country groupings points to a number of water reuse projects as parts of small-to-medium urban projects in China (BRIC), against major projects identified in Peru and Singapore (rest of the world, average capacity 522,500 m^3 per day, $n = 4$).

Water reuse PPPs and water stress

There is too much variability in renewable water resources and water demand within countries to apply national water-stress data to the development of water reuse facilities. Instead, the WaterGAP model for water stress at river-basin level in 2000 was used (Alcamo et al., 2003a, 2003b), which uses the ratio of total withdrawals to total renewable supply. Facility locations were classified as no stress (ratio < 10%), low stress (10–20%), mid stress (20–40%, high stress (40-80%), or very high stress (>80%).

The WaterGAP map as it stands does not take into account the impact of certain cities on their water basins, so Windhoek is classified as mid stress, when in fact it is in a very high-stress area. As a result, the actual bias is probably greater than indicated here.

Tables 8 and 9 outline how water stress has played a role in the development of PPP contracts. There is a pronounced bias towards contract awards taking place in areas of very high

Table 8. Water reuse PPP contract awards and water stress.

	Number of contracts	Population (millions)	Capacity (million m^3 per year)
No stress (<10%)	3	1.24	0.26
Low stress (10–20%)	3	0.13	0.05
Mid stress (20–40%)	15	5.26	1.91
High stress (40–80%)	11	2.98	3.09
Very high stress (>80%)	40	19.83	6.50
Total	72	29.44	11.80

Note: % stands for the percentage of renewable resources abstracted.

Table 9. Water reuse PPP contract starting dates in relation to water stress (millions of people).

Population served	1995–99	2000–04	2005–09	2010–14	1995–2014
No or low stress (<20%)	0.00	0.01	0.05	1.31	1.37
Mid stress (20–40%)	0.00	0.11	2.27	2.89	5.26
High stress (40–80%)	0.00	0.57	1.40	1.01	2.98
Very high stress (>80%)	0.74	2.53	9.60	6.98	19.85
Total	0.74	3.22	13.31	12.18	29.44

Note: % stands for the percentage of renewable resources abstracted.

Table 10. Water reuse PPP contracts and water scarcity compared with global population and GDP.

Water stress	Global population	Global GDP	Number of contracts	Population-served	Capacity of facilities
<20%	46%	59%	8%	5%	3%
20–40%	18%	16%	21%	18%	16%
>40%	36%	22%	71%	77%	81%

Note: % stands for the percentage of renewable resources abstracted.

Table 11. Ratio of water reuse PPPs to population and GDP, by degree of water scarcity.

Population	Contracts	Population	Capacity
<20%	1.00	1.00	1.00
20–40%	6.71	9.20	13.63
>40%	11.34	19.68	34.50
GDP	Contracts	Population	Capacity
<20%	1.00	1.00	1.00
20–40%	9.68	13.28	19.67
>40%	23.80	41.30	72.41

Note: % stands for the percentage of renewable resources abstracted.

water stress. Six contracts were awarded in areas of no or low stress, 15 for mid stress and 51 for high or very high stress. In no-stress and low-stress areas, average project capacity was 52,000 m^3 per day, against 127,000 m^3 per day and 129,000 m^3 per day in medium-, high-, and very high-stress areas, respectively.

Contracts started earliest in areas with very high water stress and developed slowest in those with no or low stress. By 2010–14, there was significant activity in all areas. Throughout the period, the majority of contract awards took place in high-stress or very high-stress areas.

The distribution of contracts is further skewed when considering population and GDP. Veolia (2011) developed a data-set for global population and GDP (base year 2010) and compared these with water scarcity. While 54% of the global population live in areas of water scarcity, these areas account for only 41% of global GDP. Table 10 compares population and GDP distribution with water reuse PPPs.

Expressing the distribution of contract awards, population served and reuse capacity in areas with medium or high stress against those in low-stress areas highlights the greater frequency of water reuse PPPs in water-scarce areas.

Table 11 shows how the ratio of contracts (by number, population served and capacity) rises in the higher water stress areas over lower stress areas. The bias towards using water reuse PPP projects increases with water scarcity. It is also evident that as a whole, water reuse PPPs are being used in the less well-off areas. The growing water reuse in China (along with a favourable attitude towards PPPs) reflects this. Contract size and frequency both rise with increasing water scarcity in both population and GDP terms (it is more pronounced in GDP terms).

Water reuse PPPs in the context of water and wastewater PPPs

Table 12 places the development of water reuse PPPs in the context of other water and wastewater PPP contract awards in the GWI database in population terms.

Comparing water treatment plant contract awards ($n = 248$) and wastewater treatment plant contract awards ($n = 975$) it is worth noting that the average contract size for water projects identified was 179,122 m^3 per day, versus 87,647 m^3 per day for wastewater. Overall,

Table 12. Water reuse in the context of all water and wastewater PPPs, 1995–2014.

	1995–99	2000–04	2005–09	2010–14	1995–2014
People served (millions)	210.2	276.0	386.2	215.2	1,087.6
Number of contracts	321	670	1,102	621	2,714
Average size (million people)	0.65	0.41	0.35	0.35	0.40
Reuse: % of all PPP contracts	0.4%	1.2%	3.4%	5.7%	2.7%
Reuse: size relation to average PPP contract	0.57	0.86	1.31	1.10	1.02

it would be expected that wastewater projects would be smaller, as not all treated water will reach a wastewater treatment plant. The fact that water reuse plants are on average 85% larger than wastewater treatment plants is likely to reflect their use as part of larger wastewater treatment projects.

In 1995–2004, water reuse was not a significant element in PPPs in population terms and its contract sizes were smaller than in the rest of the area. From 2005, contract sizes are typically at least as large as in the rest of the sector and water reuse occupied, at 2.7%, a small but significant part of the overall PPP contract market. The proportion of contracts awarded for water reuse rose consistently during the period, from 0.4% in 1994–99 to 5.7% in 2010–14, which indicates that the impact of water reuse within PPP is set to continue to increase in at least the medium term.

Looking at water and wastewater treatment projects, a narrower definition of PPPs which more closely resembles the nature of a water reuse project, 1223 such projects were awarded between 1995 and 2014, with a total capacity of 129.88 million m^3 per day, an average capacity of 106,197 m^3 per day, and a total population served of 379.15 million. Water reuse projects accounted for 7.2% of water and wastewater treatment and water reuse projects during this period in terms of people served, and 8.3% of total treatment capacity.

Are water reuse PPPs contentious?

No contracts were noted which had ended either by unilateral action (by the municipality or the government) or by a sale of the contract back to the municipality or government. Four contracts ended at their agreed expiry date, and all were subsequently renewed.

Looking at all water and wastewater PPPs, the author's Envisager PPP database currently includes 1684 contracts outside France and the US, covering 1,012.6 million people. It includes 44 contracts that were ended unilaterally, covering 50.42 million people, and 28 which were ended by negotiated sale, covering 42.65 million people. In population terms, 9.2% of PPP contracts identified in the database ended early. The fact that none of the water reuse PPP contracts identified to date has ended early would indicate that they are less contentious than other water and wastewater PPP contracts.

Policy considerations

Where water is provided at a low price, economic incentives for water reuse are minimized, which in turn makes it less attractive for PPPs. Legislation can have an impact here. For example, in the European Union, the effective enforcement of the EU Water Framework Directive (2000/60/EC) would see the adoption of full cost recovery pricing for water provision, while the Urban Wastewater Treatment Directive (1991/271/EC) encourages water reuse "where appropriate".

Another incentive for water reuse PPPs will be commonly accepted operational standards. The International Organization for Standardization is currently drawing up standards for water reuse. The ISO 16075 Guidelines for Treated Wastewater Use for Irrigation Projects are under development by ISO/TC 282 (Standardisation of Water Re-use of Any Kind and for Any Purpose). Meanwhile, ISO 24510 (Activities Relating to Drinking Water and Wastewater Services) has been published, and outputs relevant to water reuse are being developed by ISO/TC 224. ISO 24510 includes a proposed series of standards for the performance of water and wastewater systems that are applicable to water reuse.

Discussion

The small size of the database as it stands means that the outcomes are indicative and have not been subjected to statistical analysis. A number of outcomes and trends are potentially significant, especially in relation to the bias of contract awards towards areas of greater water scarcity.

This means that the data presented here are best seen as an overview of what has happened to date in this area and to serve as a basis for developing a more analytical approach as further contracts are announced over time. The prime motive here is to highlight the potential that this area has to offer as and when further information emerges. As the frequency of contract awards rises, more analytical approaches will be possible. According to GWI (2015), as of November 2015 there were 199 water reuse PPP project proposals at various stages of development.

Identifying and following water and wastewater contracts is not easy because availability of information is poor. The higher profile enjoyed by PPP contracts means that a more accurate picture of their development can be obtained than for municipally operated facilities.

No water reuse PPP contracts have ended before their expiry date. This is in marked contrast with water and wastewater PPP contracts in general. This has the potential to form part of a more nuanced approach towards the role of PPP in water services than has typically been the case.

PPPs for water reuse projects are set to play a significant role in adapting to current and future water scarcity. The areas which currently have the greatest water scarcity have a lower GDP per capita than areas with low water scarcity. As these economies develop, per capita water use is expected to rise, redoubling the need for new water resources.

As with PPPs in general, water reuse PPP projects are not a universally applicable way of addressing current or future water supply needs; rather, this is an approach that under the right circumstances can offer the combination of cost and performance benefits municipal and industrial clients require.

Future areas for research

There is the prospect of greater granularity for matching projects with water resources as new surveys are developed, along with considering water usage by municipal, industrial and agricultural perspectives.

Water reuse is increasingly being considered in less water-scarce areas. There is the potential to consider the deployment of water reuse PPPs in relation to planning for increased future water scarcity.

More data and a longer time scale will allow formal statistical analysis. This may also enable the impact of demographic change over a period of time to be investigated, especially in considering when areas are classified as more water-scarce.

Obtaining data on public-sector water reuse projects would allow analysis of the relationship between public and private management of water reuse projects. This in turn could be applied to the impact of water reuse on water demand management at the regional or water basin level.

The impact of wastewater treatment PPP contracts being developed to include water recovery after the contract has started has not been considered. Anecdotal evidence suggests that this is becoming increasingly significant.

Where comparable data are available, the costs of developing PPP and municipal water reuse projects could be compared, in terms of both capital and operating costs.

Disclosure statement

No potential conflict of interest was reported by the author.

References

Alcamo, J., Döll, P., Henrichs, T., Kaspar, F., Lehner, B., Rösch, T, & Siebert, S. (2003a). Development and testing of the WaterGAP 2 global model of water use and availability. *Hydrological Sciences Journal, 48*, 317–337.

Alcamo, J., Döll, P., Henrichs, T., Kaspar, F., Lehner, B., Rösch, T, & Siebert, S. (2003b). Global estimation of water withdrawals and availability under current and "business as usual" conditions. *Hydrological Sciences Journal, 48*, 339–348.

Bakker, K. (2004). *An uncooperative commodity*. Oxford, UK: Oxford University Press.

Bakker, K. (2010). *Privatizing water: Governance failure and the world's water crisis*. Ithaca, USA: Cornell University Press.

Chen, W., Bai, Y., Zhang, W., Lyu, S., & Jiao, W. (2015). Perceptions of different stakeholders on reclaimed water reuse: The Case of Beijing, China. *Sustainability, 7*, 9696–9710.

GWI. (2015). Project tracker, global water intelligence, 46–53 GWI, November 2015.

Hartley, T. W. (2006). Public perception and participation in water reuse. *Desalination, 187*, 115–126.

Jimenez, B., Asano, T., Ellis, B., Bertrand-Krajewski, J-L., Binnie, C. & Kimber, M. (2008). *Water reuse - An International Survey of current practice, issues and needs*. London, UK: IWA Publishing.

Keremane, G. B. & McKay, J. (2009). Critical Success Factors (CSFs) for private sector involvement in wastewater management: the Willunga Pipeline case study. *Desalination, 244*, 248–260.

Lahnsteiner, J. & Lempert, G. (2007). Water management in Windhoek. *Water Science & Technology, 55*, 441–448.

Lazarova, V., et al. (2013). *Milestones in water reuse: The best success stories*. London, UK: IWA Publishing.

MAL. (2014). *Global water markets 2015*. Oxford, UK: Media Analytics Limited.

MAL. (2015). *Private sector participation in water*. Oxford, UK: Media Analytics Limited.

Molinos-Senante, M., Hernandez-Sancho, F., & Sala-Garrido, R. (2013) Tariffs and cost recovery in water reuse. *Water Resources Management, 27*, 1797–1808.

Owen, L. (2015). *inDepth: The Arup Water Yearbook, 2015–16*. London, UK: Arup.

PUB. (2014). A Fine Balance. Annual Report 2013/2014, Singapore Public Utilities Board, Singapore

Veolia. (2011). *Finding the blue path for a sustainable economy*. Chicago, USA: Veolia Water North America.

The regulatory framework of reclaimed wastewater for potable reuse in the United States

Rosario Sanchez-Flores[a], Adam Conner[b] and Ronald A. Kaiser[a]

[a]Water Management and Hydrological Sciences, Texas A&M University, College Station, USA; [b]Water Resources, San Antonio Water System, TX, USA

ABSTRACT

Water scarcity, climate change, population growth and rising infrastructure costs have opened the door for unconventional or 'new' water sources. Reclaimed water reuse has historically been practised for potable use in the United States as de facto water reuse or unplanned indirect water reuse. The increasing number of planned indirect water reuse projects in the country and the approval of the first direct potable reuse projects have exposed the limitations of the regulatory system at the national and state levels. These limitations pose barriers and/or add uncertainty to the viability of potable water reuse.

Introduction

As water scarcity increases and population projections continue to trend upward in the United States, the need to stretch the available water resources and to develop 'new' or 'unconventional' water resources in some regions has become the highest priority in water management. The reuse of wastewater for indirect or direct potable use is a 'new' (Meehan, Ormerod, & Moore, 2013) or alternative water source that has recently become a necessary alternative for several regions in the US. Direct potable reuse (DPR) refers to the use of reclaimed water that is piped directly from an advanced wastewater treatment facility to the drinking water treatment facility or distribution system, whereas indirect potable reuse (IPR) typically involves placing the treated effluent back into water supply sources, such as rivers, lakes or aquifers, to be retrieved later for potable use. Although this practice has not been uniformly implemented – or accepted – across the country, its use has dramatically increased in the Southwest and in certain locations in the eastern part of the country (primarily Virginia and Florida).

As water scarcity worsens due to climate change and population growth (particularly in coastal and dryland regions), the associated economic losses, public health risks, environmental degradation and social and even political pressure are affecting societies in an unprecedented way. The legal framework for wastewater or reclaimed water reuse for potable uses (primarily direct reuse) is equally uneven and non-standardized. The federal government has developed general information and guidelines for water reuse since 1992, although enforcement is limited to the regulations established under the Clean Water Act of 1972, the Safe Drinking Water Act of 1974 and following amendments. State guidelines for water reuse, including water quality standards, vary greatly among and within states, and only a few states have established regulations. Different water ownership doctrines between states also play a significant role in defining the extent of and potential for water reuse. Recent federal protocols and research have just started to open discussions regarding the development of federally standardized regulations and guidelines, specifically for water reuse for potable uses (EPA, 2012; NRC, 2012). As the traditional paradigm begins to show its limits and new reservoir construction projects have reached their ecological and beneficial offsets as traditional water suppliers in some areas (Kenny et al., 2009), considerations of technological reliability, economic feasibility and social acceptability must be analyzed as a new paradigm of urban water management evolves (Meehan et al., 2013).

The actual legal, technical and scientific framework for water reuse, specifically for potable use, deserves special attention. The population in the US has doubled since the 1950s, while municipal water supplies have tripled, and these changes have become increasingly concentrated in urban regions (NRC, 2012). In terms of water demand, municipal use remains a small fraction of the total relative to agricultural and energy demands, but the demand curve for public water supply is expected to exhibit the greatest increase in the coming decades (Gleick, 2003; Kenny et al., 2009). Water reuse as a water augmentation strategy for potable use will require a more structured and standardized framework to guarantee future water availability for urban populations. Improving water quality as a means to increase total water availability through water reuse requires continued advancements in wastewater and drinking water treatment techniques, monitoring systems, standardization of regulations, social involvement, proactive and preventive measures, and conservation programmes. These challenges must be resolved in the short term, because water demand has already surpassed water supplies in some rapidly growing regions of the US, particularly in the Southwest.

This article reviews the present legal framework governing treated wastewater or reclaimed water reuse for potable uses and addresses the differences between IPR, either planned or unplanned (de facto) reuse (Asano, Burton, Leverenz, Tsuchihashi, & Tchobanoglous, 2007), and DPR, specifically in the US. This study covers the topic of water ownership and its relevance to IPR and DPR. Secondly, this review evaluates the potential for water reuse in the country and drawbacks and challenges for IPR and DPR. It covers specific federal regulations that govern the topic and their implications for water quality standards and requirements, and how states differ in terms of water ownership under state law. Lastly, the article highlights water reuse case studies that deserve mention because they are representative of IPR and DPR cases in the US and internationally.

Background and present conditions of IPR, de facto reuse and DPR in the US

Nationally, approximately 12 billion gallons of wastewater are discharged into estuaries and the ocean daily. The discharge is equivalent to 27% of the total water use of public supplies in the US (NRC, 2012). If this water were reused, the additional water availability would represent nearly 30% of the municipal water use in 2005 (Kenny et al., 2009). Currently, the states of California and Florida have the largest water reuse programmes (for both potable and nonpotable applications), with reuse of approximately 500 million gallons per day (MGD), followed by the states of Texas and Arizona, with an average of 200 MGD (Rock, McLain, & Gerrity, 2012). These four states account for 90% of the total water reuse in the country, although other states, including Washington, Oregon, Nevada, Colorado, New Mexico, Pennsylvania, Virginia, Georgia and Maryland, have recently begun to develop more intensive reuse programmes (Rock et al., 2012; WateReuse Association, 2011).

The potential for water reuse to augment potable water is significant; however, the practice has been generally limited to planned IPR and/or de facto potable reuse (unplanned potable water reuse) because state, federal and city agencies tend to consider DPR projects an option of last resort (NRC, 2012). IPR can be defined as the augmentation of drinking water supplies with purified or treated water using an environmental buffer (e.g. groundwater recharge, direct injection, soil filtration, basin spreading, blending or surface augmentation; NRC, 2012). One of the earliest IPR projects in the country began development in 1971: the Upper Occoquan Service Authority in Northern Virginia returns treated water to a reservoir (Upper Occoquan Service Authority, 2014). In contrast, unplanned IPR, or de facto potable reuse, is a regular practice in the US and has been commonly implemented for at least 100 years (Tchobanoglous et al., 2015; Vigneswaran & Sundaravadivel, 2004). By the 1940s, approximately half of the US urban population was served with treated wastewater (NRC, 2012). 'De facto reuse', as it was originally called (Asano et al., 2007), refers to those projects that incorporate water reuse but are not officially recognized as permitted reuse projects. In any case, the volume of unplanned IPR or de facto potable reuse has been growing steadily in the country, as treated effluent discharges into rivers and streams and is later reused downstream (NRC, 2012). According to the US Environmental Protection Agency (EPA), Region 6, approximately 60% of the surface waters in the region that includes the states of Arkansas, Texas, Louisiana, New Mexico and Oklahoma receives at least 10% treated effluent under low-flow conditions (Brooks, Riley, & Taylor, 2006). Large cities such as Philadelphia, Nashville, Cincinnati, New Orleans and Houston are examples of de facto potable reuse because their drinking supplies come from the waters of the Delaware, Cumberland, Ohio, Mississippi and Trinity River, respectively (EPA, 2012). Because de facto potable reuse has been a reality for many decades, technological research and development to improve the performance of water treatment plants to comply with water quality standards has been continuously evolving, and in some cases, their performance has exceeded environmental treatment options. The Orange County Water District's Montebello Forebay in California, the Occoquan Reservoir in Virginia and the El Paso Water Utility's Hueco Bolson are examples of IPR projects that have been able to meet purification standards for drinking water and have gained public acceptance (EPA, 2012).

Despite this trend, the City Council of San Diego, California, was able to gain approval for the implementation of a large-scale IPR project only in 2014, as a result of the persistent drought conditions in the state. Similarly, in the state of Texas, the Colorado River Municipal Water District (headquartered in Big Spring) and the city of Wichita Falls were the first DPR projects to begin operation in the nation (although Wichita Falls has since transitioned to IPR after significant rainfall). Given the lack of standardized regulations on IPR and DPR at the national scale, the Texas Commission on Environmental Quality decided to approve these projects on a case-by-case basis (Steinle-Darling, 2015). Documented temporary cases, such as the Pure Cycle Corporation in Colorado, have shown the potential of DPR, but financial constraints and public rejection prevented it from continuing operation (Pure Cycle Corporation) or delayed their operation (Villa of Cloudcroft) (Tchobanoglous, Leverenz, Nellor, & Crook, 2011).

The city of Chanute, Kansas, in the 1950s used treated wastewater as a primary drinking supply for five months when the flow of the main source of public supply, the Nesho River, dropped too low (Vigneswaran & Sundaravadivel, 2004). The Denver Potable Reuse Demonstration Project in 1984 was intended to reuse wastewater for potable use but faced strong opposition and financial constraints (Tchobanoglous et al., 2011).

DPR, in which purified municipal wastewater is introduced directly into the water distribution system, is only now becoming attractive for water augmentation in the US (Tchobanoglous et al., 2011). This increasing interest is being driven by considerations related to water rights. Allocated water can be lost through numerous avenues, including release of the wastewater downstream or into reservoirs and lakes; evaporative losses; unsuitable IPR conditions, such as a lack of environmental buffers close to the wastewater treatment plant; potential contamination once water is released into the buffer; and the maintenance, operation and monitoring costs of the environmental buffers (NRC, 2012). However, it must be said that actual and temporary DPR projects in the US have been driven primarily by extreme water scarcity, in which the cost of implementing the DPR project was lower than that of the alternatives.

Potential for DPR in the US

The primary difference between IPR and DPR is the relationship between the treatment facility and the drinking distribution system. IPR assumes that an environmental buffer acts as a catalyst for water purification before the water is returned to the drinking distribution system. The rationale for DPR is that wastewater is sent back directly to the drinking distribution system after advanced treatment at a highly regulated and monitored facility. Expecting improved water quality to occur by placing already very highly treated water into a surface water body is highly questionable. There is no evidence that environmental buffers are more effective in purifying water than certain advanced engineering systems, and some engineering systems in fact might be even more effective than natural systems (EPA, 2012; NRC, 2012). Though DPR has gained importance in recent years, mostly in the Southwest, there is still some public opposition to this practice, and the major emphasis for wastewater reuse to date has been on nonpotable uses. The primary obstacles to the acceptance and promotion of DPR projects by agencies in the US are considered to be the lack of historical assessments of the proportion of wastewater effluent that has actually been reused for potable uses (NRC, 2012; Rock et al., 2012) and the potential health risks associated with uncertainties in the reliability of treatment systems that produce purified water that meets all of the required drinking standards (EPA, 2012; Miller, 2015). Similarly, the 'yuck factor', i.e., the disgust associated with the *idea* of drinking wastewater, still governs the decision making of water augmentation strategies (Ormerod & Scott, 2012). On the other hand, water treatment technology has been improving over the last 35 years and has been able to prove its efficiency and safety in meeting drinking water standards for 'unconventional' water resources (Cotruvo, 2015; Tchobanoglous et al., 2015). Therefore, the greater challenge does not appear to be technological, but more at the managerial and public policy level.

Planned water reuse for potable uses is estimated to represent less than 1% of the total water supply in the country (NRC, 2012). Recommendations to enhance the performance of treatment systems for an optimal and reliable DPR system include the enhancement of source control programmes, physical screening, upstream flow equalization, elimination of untreated return flows, improved performance of monitoring systems, modification of bio-logical treatment, and the development of pilot test facilities to evaluate new technologies (Tchobanoglous et al., 2011). Unfortunately, these recommendations imply investment and considerably higher budgets, which are not necessarily cost-effective for all water treatment facilities in the country. In addition, the recent emphasis on 'contaminants of emerging con-cern' may lead to the overdesign of water treatment facilities, increasing the marginal cost per acre-foot (Gerrity, Pecson, Trusell, & Trusell, 2013). Although it has been recognized that climate, geology and site-specific conditions play important roles in defining the site-specific costs of water treatment systems, DPR projects tend to be generally more expensive than most water conservation alternatives (NRC, 2012). Likewise, low potable water rates usually do not encourage an investment in potable water reuse unless water scarcity is such that potable water reuse is the most cost-efficient alternative (Western Resource Advocates, 2012).

However, research has suggested that the benefits derived from DPR projects should also be valued from a nonmonetary perspective, considering the advantages in terms of energy conservation and environmental preservation. The reduction in water importation or water transfers, infrastructure construction (including reservoirs and canals) and pumping costs, and the improvements in water quality and water savings from conservation, can also be equally significant (Gerrity et al., 2013; NRC, 2012; Schroeder, Tchobanoglous, Leverenz, & Asano, 2012). The two DPR cases in the US, though very recent and pursued for different reasons (Wichita Falls was approved as an emergency response and has since transitioned back to IPR; Steinle-Darling, 2015), seem to be signs of social acceptance. Their successful implementation has the potential to inspire a long-term movement that will be attractive in other water-scarce regions in the state, the country and eventually the world (Martin, 2014). Likewise, the stigma once established by the National Research Council in their 1998 report, in which it was recommended that the use of reclaimed water for potable purposes be an "option of last resort" (NRC, 1998), has been replaced in their 2012 version, which states that potable reuse should be considered not a last-resort option but a real alternative for future water demand (Miller, 2015; NRC, 2012).

Federal regulations applicable to IPR and DPR

At the federal level, two primary laws govern water quality in the US. The first is the Clean Water Act of 1972, which is administered by the Environmental Protection Agency (EPA). The main purpose of this act is to give federal and state governments the responsibility of guaranteeing that the water quality of all public surface waters is *fishable and swimmable* by protecting the chemical, physical and biological integrity of the nation's waters through the regulation of wastewater discharges from agricultural, industrial and municipal pollu-tion point sources (EPA, 2012). However, overland (non-point source) flow from agricultural practices and irrigation return flows is not directly regulated by the Clean Water Act, which also does not address groundwater pollution. The National Pollutant Discharge Elimination System (NPDES), which is mandated by the Clean Water Act, was designed to protect down-stream users from untreated effluents that could potentially be used for potable uses. NPDES permits target a broad range of ordinary pollutants (biochemical oxygen demand, total suspended solids, faecal coliforms, oil and grease, and pH) and a list of 126 priority pollut-ants, but it does not cover every pollutant, especially those of recent concern such as total

Table 1. Guidelines for IPR suggested in the EPA guidelines for water reuse.

Reuse category and description	Treatment	Reclaimed water quality[1]	Reclaimed water monitoring	Setback distances[2]	Comments
Groundwater recharge by spreading into potable aquifers	Secondary[3] Filtration[4] Disinfection[5] Soil aquifer treatment	Including but not limited to: • No detectable total coliform/100 mL[8,9] • 1 mg/L Cl2 residual (min.)[10] • pH = 6.5–8.5 • ≤ 2 NTU[7] • ≤2 mg/L TOC of wastewater origin • Meet drinking water standards after percolation through vadose zone	Including but not limited to: • pH – daily • Total coliform – daily • Cl2 residual – continuous • Drinking water standards – quarterly • Other[15] – depends on constituents • TOC – weekly • Turbidity – continuous • (Monitoring is not required for viruses and parasites)	Distance to nearest potable water extraction well that provides a minimum of 2 months retention time in the underground.	• Depth to groundwater should be at least 6 feet at the maximum groundwater mounding point. • The reclaimed water should be retained underground for at least 2 months prior to withdrawal. • Monitoring wells are necessary to detect the influence of the recharge operation on the groundwater. • Reclaimed water should not contain measurable levels of pathogens after percolation through the vadose zone.[11] • Recommended log-reductions of viruses, Giardia, and Cryptosporidium can be based on challenge tests or the sum of log-removal credits allowed for individual treatment processes. Monitoring for these pathogens is not required. • Dilution of reclaimed water with waters of non-wastewater origin can be used to help meet the suggested TOC limit.
Groundwater recharge by injection into potable aquifers	Secondary[3] Filtration[4] Disinfection[5] Advanced wastewater treatment[14]	Including but not limited to: • No detectable total coliform/100 mL[9,10] • 1 mg/L Cl2 residual (min.) • pH = 6.5–8.5 • ≤2 NTU[7] • ≤2 mg/L TOC of wastewater origin • Meet drinking water standards	Same as above	Same as above	• The reclaimed water should be retained underground for at least 2 months prior to withdrawal. • Monitoring wells are necessary to detect the influence of the recharge operation on the groundwater. • Recommended quality limits should be met at the point of injection. • The reclaimed water should not contain measurable levels of pathogens at the point of injection. • Higher chlorine residual and/or a longer contact time may be necessary to assure virus inactivation. • Recommended log-reductions of viruses, Giardia, and Cryptosporidium can be based on challenge tests or the sum of log-removal credits allowed for individual treatment processes. Monitoring for these pathogens is not required. • Dilution of reclaimed water with waters of non-wastewater origin can be used to help meet the suggested TOC limit.

| Augmentation of surface water supply reservoirs | Secondary[3] Filtration[4] Disinfection[5] Advanced wastewater treatment[14] | Same as above | Site specific – based on providing 2 months retention time between introduction of reclaimed water into a raw water supply reservoir and the intake to a potable water treatment plant. | • The reclaimed water should not contain measurable levels of pathogens.[11]
• Recommended level of treatment is site-specific and depends on factors such as receiving water quality, time and distance to point of withdrawal, dilution and subsequent treatment prior to distribution for potable uses.
• Higher chlorine residual and/or a longer contact time may be necessary to assure virus and protozoa inactivation.
• Recommended log-reductions of viruses, Giardia, and Cryptosporidium can be based on challenge tests or the sum of log-removal credits allowed for individual treatment processes. Monitoring for these pathogens is not required.
• Dilution of reclaimed water with water of non-wastewater origin can be used to help meet the suggested TOC limit. |

TOC = total organic carbon; NTU = nephelometric turbidity units

(1) Recommended quality limits apply to the reclaimed water at the point of discharge from the treatment facility.[t/s: The notes start with no. 2—please renumber]

(2) Setback distances are recommended to protect potable water supply sources from contamination and to protect humans from unreasonable health risks due to exposure to reclaimed water.

(3) Secondary treatment should produce effluent in which both BOD and SS do not exceed 30 mg/L.

(4) Filtration means the passing of wastewater through natural undisturbed soils, filter media such as sand or anthracite, or microfilters.

(5) Disinfection means the destruction, inactivation, or removal of pathogenic microorganisms by chemical, physical, or biological means.

(6) As determined from the 5-day BOD test.

(7) The recommended turbidity should be met prior to disinfection. The turbidity should not exceed 5 NTU at any time. If SS is used in lieu of turbidity, the average SS should not exceed 5 mg/L. If membranes are used as the filtration process, the turbidity should not exceed 0.2 NTU and the average SS should not exceed 0.5 mg/L.

(8) Recommended coliform limits are median values determined from the bacteriological results of the last 7 days for which analyses have been completed. Either the membrane filter or the fermentation tube technique may be used.

(9) The number of total or faecal coliform organisms should not exceed 14/100 ml in any sample.

(10) This recommendation applies only when chlorine is used as the primary disinfectant. The total chlorine residual should be met after a minimum actual modal contact time of at least 90 min., unless a lesser contact time has been demonstrated to provide indicator organism and pathogen reduction equivalent to those suggested in these guidelines. In no case should the actual contact time be less than 30 min.

(11) It is advisable to fully characterize the microbiological quality of the reclaimed water prior to implementation of a reuse programme.

(12) The number of faecal coliform organisms should not exceed 800 per 100 ml in any sample.

(13) Some stabilization pond systems may be able to meet this coliform limit without disinfection.

(14) Advanced wastewater treatment processes include chemical clarification, carbon adsorption, reverse osmosis and other membrane processes, advanced oxidation, air stripping, ultrafiltration and ion exchange.

Adapted from (EPA, 2012).

organic carbon (TOC), oxyhalide and microbial pathogens (Gerrity et al., 2013). In addition, the NPDES programme has established limits on the mass concentrations of specific contaminants (known as total maximum daily load, or TMDL) that are discharged into receiving water bodies (EPA, 2012). The NPDES programme is significant because almost two-thirds of the potable water delivered by public water systems in the country comes from surface waters, such as reservoirs, rivers and lakes (NRC, 2012). For example, the Trinity River in Texas, which is the most important water source for the city of Houston, largely contains wastewater effluent from the Dallas–Fort Worth area under base-flow conditions. In 2002, the Trinity River Authority established one of the first large-scale IPR projects to divert wastewater into engineered wetlands before discharging downstream. Although the Clean Water Act also encourages water reclamation or reuse, it is not required (EPA, 2012). States may choose not to use the NPDES programme to regulate their water reuse water quality standards at a local level.

The second federal law governing water quality is the Safe Drinking Water Act (SDWA), which was approved in 1974. The SDWA has the responsibility of setting maximum contaminant levels (MCLs) or treatment requirements for all of the country's drinking water. The act grants authority to the EPA to establish and enforce national health-based water quality standards for drinking water to protect against both naturally occurring and man-made contaminants. It also requires that water systems test for the regulated contaminants to treat or remove the contaminants if they exceed the MCLs. These standards affect the degree to which reclaimed water can be used for potable water reuse. Specifically for water reuse, the EPA has developed a compilation of guidelines intended to serve as a reference for water reuse, particularly in states where criteria have not been established. However, each state and city in the US is deemed responsible for its own reuse policy as long as the federal regulations related to the Clean Water Act and SDWA are followed. Federal guidelines are mainly directed to protect public health by setting standards to control pathogenic microorganisms for non-potable reuse and health-significant microorganisms and chemical contaminants for all public drinking water supplies regardless of source (EPA, 2012). The maximum MCLs recommended by the EPA for the nutrients nitrate and nitrite are 10 mg/L and 1 mg/L, respectively (NRC, 2012). The maximum recommended EPA guideline for TOC, a measure of the organic content in wastewater, was reduced from 4 mg/L in 2004 to 2 mg/L in 2012 (EPA, 2012). At the local level, state regulators can set their own water quality standards which must be no less stringent than federal standards. For example, the state of California has adopted IPR via groundwater recharge and set the standard for TOC at 0.5 mg/L (EPA, 2012). For drinking water, federal standards require no detectable coliform per 100 mL as the most commonly used indicator of disinfection efficiency. However, studies have shown that this standard is not the best indicator for tracing parasites, particularly *Giardia* and *Cryptosporidium*, and other contaminants of emerging concern, such as *E. coli* (NRC, 2012).

Present federal guidelines suggest appropriate disinfection processes that are able to reduce or inactivate enteroviruses to low and acceptable levels (EPA, 2012). Some states, including California, have already established enforceable regulations setting the limit for perchlorate at 6 µg/L, whereas the state of Massachusetts has a stricter enforceable limit of 2 µg/L (NRC, 2012). In 1989, the EPA approved the Surface Water Treatment Rule under the SDWA with the purpose of offering guidance to water treatment plants on disinfection levels for pathogens (*Legionella* and *Giardia*). The rule provides a benchmark for disinfection practices for viruses and bacteria to guarantee a maximum of a 1-in-10,000 per capita risk of infection (EPA, 2010). It also recommends the development of a 'hazard analysis and critical control points' framework for application at the state and city levels to improve the monitoring, control and standardization of treatment techniques to assure a confident level of operational reliability of water treatment plants (NRC, 2012). The SDWA requirements apply to more than 170,000 public and private water systems that supply drinking water to the public in the US (EPA, 2015). Table 1 shows the specific federal guidelines for IPR practices and water quality standards in the US.

Overall, the recognized challenge related to water quality standards for IPR and DPR projects in the US is not the detection of contamination but the identification, prevention and control of environmental impacts and health risks that are associated with exposure to contaminants (NRC, 2012). To date, no regulations or criteria have been proposed for DPR in the US (EPA, 2012), and because the current DPR projects in the country have been approved on a case-by-case basis, most of the potable reuse projects in the country are considered IPR (Gerrity et al., 2013).

Overview of states' regulations and guidelines applicable to IPR and DPR

Even though all states follow drinking water regulations or guidelines that apply to IPR and DPR, only a limited number of states in the country have regulations or guidelines specifically for IPR. The difference between regulations and guidelines is enforceability: regulations are enforceable, whereas guidelines are recommendations with limited enforceability. Where regulations or guidelines are available, they include requirements for treatment, water quality standards and monitoring. Water quality standards usually include limits on total suspended solids (TSS), nutrients, TOC, turbidity and total coliform. Generally, states also specify the minimum time that reclaimed water must be retained in the environmental buffer and the required distance between the recharge and withdrawal points (EPA, 2012). Treated effluent is commonly disinfected using chlorination, ultraviolet light and ozone, but treatments vary depending on local regulations.

Conventional drinking water treatment plants generally follow the standard 4-log virus removal, 3-log *Giardia* removal and between 2-log and 5.5-log removal of *Cryptosporidium* (Steinle-Darling, 2015). Table 2 shows a snapshot of water quality standards for IPR in four selected states where IPR projects have been successfully implemented. The table highlights the states of California, Florida, Texas and Washington, as they have developed more specific water quality criteria for specific treatments. Other states, such as Hawaii and Virginia, have developed water quality criteria only on a case-by-case basis. Arizona, New Jersey, Nevada and North Carolina have yet to develop state-level regulations or specific criteria for different types of IPR. In the case of Arizona, all state requirements are related to groundwater recharge in nonpotable aquifers (EPA, 2012).

The states that have had the most successful experiences with water reuse projects and the development of appropriate regulations and guidelines are Arizona, California, Florida, Hawaii, Nevada, New Jersey, Pennsylvania, North Carolina, Texas, Massachusetts, Utah, Virginia and Washington. They all have different categories of water reuse but similar water quality requirements for different types of use. Among these states, only California, Florida, Arizona, Texas, Massachusetts, Washington, Utah and Virginia have regulations specifically for IPR (EPA, 2012).

Table 2. Indirect potable reuse water quality standards in selected states.

				Washington		
	California[1]	Florida[3]	Texas	Surface percolation, Class A	Direct groundwater recharge,[4] Class A	Streamflow augmentation, case-by-case
Treatment (system design) requirements						
Unit processes	Oxidized, coagulated, filtered, disinfected, multiple barriers for pathogen and organics removal	Secondary treatment, filtration, high-level disinfection, multiple barriers for pathogen and organics removal	Case-by-case	Oxidized with nitrogen reduction, filtered, disinfected	Oxidized, coagulated, filtered, RO-treated, disinfected	Oxidized, clarified, disinfected
UV dose, if UV disinfection used	NWRI (2012) guidelines[2]	NWRI (2012) UV guidelines enforced, variance allowed	NS	NWRI (2012) guidelines	NWRI (2012) guidelines	NWRI (2012) guidelines
Chlorine disinfection requirements, if used	CrT > 450 mg·min/L; 90 min modal contact time at peak dry weather flow[3]	TRC > 1 mg/L; 15 min contact time at peak hour flow	NS	Chlorine residual > 1 mg/L; 30 min contact time at peak hour flow	Chlorine residual > 1 mg/L; 30 min contact time at peak hour flow	Chlorine residual to comply with NPDES permit
Monitored Reclaimed Water Quality Requirements						
BOD5 (CBODs)	NS	CBOD5: −20 mg/L (ann. avg.) −30 mg/L (mon. avg.) −45 mg/L (wk. avg.) −60 mg/L (max.)	5 mg/L	30 mg/L	5 mg/L	30 mg/L
TSS	NS	5 mg/L (max)	NS	30 mg/L	5 mg/L	30 mg/L
Turbidity	−2 NTU (avg.) for media filters −10 NTU (max.) for media filters −0.2 NTU (avg.) for membrane filters −0.5 NTU (max.) for membrane filters	Case-by-case (generally 2–2.5 NTU); Florida requires continuous on-line monitoring of turbidity as indicator for TSS	3 NTU	−2 NTU (avg.) −5 NTU (max.)	−0.1 NTU (avg.) −0.5 NTU (max)	NS

Bacterial indicators	Total coliform: –2.2/100 mL (7-day med.) –23/100 mL (not more than one sample exceeds this value in 30 days) –240/100 mL (max.)	Total coliform: –4/100 mL (max.)	Faecal coliform or E. coli –20/100 mL(30-day geom.) –75/100 mL (max.) Enterococci –4/100 mL (30-day geom.) –9/100 mL (max.)	Total coliform: –2.2/100 (7-day med.) –23/100 (max.)	Total coliform: –1/100 mL (avg.) –5/100 mL (max.)	Faecal coliform: –200/100 mL (avg.) –400/100 mL (max. wk.)
Total nitrogen	10 mg/L (avg. of 4 consecutive samples)	10 mg/L (ann. avg.)	NS	NA	10 mg/L	NPDES requirements to receiving stream
TOC	0.5 mg/L	–3 mg/L (mon. avg.) –5 mg/L (max.);TOX6: <0.2 (mon. avg.) or 0.3 mg/L(max.); alternate limits allowed	NS	NA	1 mg/L	NS
Primary and secondary drinking water standards	Compliance with most primary and secondary	Compliance with most primary and secondary	NS	Compliance with SDWA MCLs	Compliance with most primary and secondary	NPDES requirements to receiving stream
Pathogens	TR	Giardia. Cryptosporidium sampling quarterly	NS	NS	NS	NS

BOD = biochemical oxygen demand; CBOD = carbonaceous BOD; TSS = total suspended solids; NS = not specified by the state's reuse regulation; NR = not regulated by the state under the reuse programme; ND = regulations have not been developed for this type of reuse; TR = monitoring is not required but virus removal rates are prescribed by treatment requirements; RO = reverse osmosis; CrT = chlorine disinfection efficacy (product of the total chlorine residual times the contact time); TRC = total residual chlorine; NTU = nephelometric turbidity units; NPDES = national pollutant discharge elimination system.

[1] CDPH Draft Regulations for Groundwater Replenishment with Recycled Water adopted in June 2014.

[2] Additional pathogen removal is required for groundwater recharge through other treatment processes in order to achieve 12-log enteric virus reduction, 10-log Giardia cyst reduction, and 10-log Cryptosporidium oocysts reduction.

[3] Florida requirements are for the planned use of reclaimed water to augment Class F-I, G-I or G-II groundwaters (US drinking water sources) with a background total dissolved solids of 3,000 mg/L or less. The reclaimed water must meet primary and secondary drinking water standards, except for asbestos, prior to discharge. The TOX limit does not apply and a total nitrogen limit is based on the surface water quality. Total organic halides (TOX) are regulated in Florida.

[4] Washington requires the minimum horizontal separation distance between the point of direct recharge and point of withdrawal as a source of drinking water supply to be 2,000 feet (610 m) and water must be retained underground for a minimum of 12 months prior to being withdrawn for drinking water supply.

Adapted from (EPA, 2012).

In the north-eastern US, regional water reuse has been growing slowly, and there are a limited number of water reuse projects. Only the states of Massachusetts, New Jersey and Vermont have water reuse regulations, and only limited data are available regarding water quality criteria for each project (EPA, 2012).

In the Mid-Atlantic region, only Delaware and Virginia have regulations on water reuse, while Pennsylvania and Maryland have only guidelines, most of which address irrigation systems. Washington, D.C., and West Virginia do not have regulations or guidelines for water reuse (EPA, 2012).

The south-eastern part of the country is considered one of the most populated and fastest-growing regions, but only the states of Mississippi, North Carolina and South Carolina have established regulations, and these vary depending on the priorities of the reuse. Florida, which has one of the most advanced water reuse schemes in the country, has established corresponding regulations for specific water uses. This has created an environment in which approximately one-half of the treated wastewater in Florida is reclaimed or reused, the highest percentage of any state (FDEP, 2012).

In the Midwest and Great Lakes regions, where states rely primarily on the nonconsumptive use of water from the Mississippi, Missouri and Ohio Rivers for mainly thermoelectric use, the utilized waters are not the best candidates for reuse because they do not replace consumed water. Nevertheless, regulations on water reuse have been evolving in the states of Illinois, Indiana, Iowa, Michigan, Missouri and Nebraska. In contrast, Kansas, Minnesota and Ohio have only limited guidelines, and Wisconsin does not have regulations or guidelines (EPA, 2012).

The South Central region is another centre of population growth in the country. Irrigation is the largest consumer of water in the region, and reclaimed or treated water is generally reused. Oklahoma and Texas have adopted water reuse regulations (excluding IPR projects), whereas New Mexico has only developed guidelines. Arkansas has only adopted water reuse guidelines, and Louisiana does not have any provisions on the subject (EPA, 2012). In the case of Texas, which is the only state in the country that has DPR projects in operation, the Texas Commission on Environmental Quality has established regulations on a case-by-case basis. The commission requires that DPR projects follow the standards set by the Surface Water Treatment Rule to achieve 4-log virus removal and 3-log removal for both *Giardia* and *Cryptosporidium*. It also encourages the monitoring of unregulated constituents (pharmaceuticals and personal care products; Steinle-Darling, 2015).

With the exception of Wyoming, states in the Mountain and Plains regions, where irrigation is the main user of water, have regulations on water reuse. The state of North Dakota is limited to guidelines that are applied on a case-by-case basis.

Contrastingly, in the fast-growing and arid Pacific Southwest, the state of Arizona has water reuse regulations, including IPR regulations, that have evolved since the 1970s, and the state recently established the Steering Committee on Arizona Potable Reuse, whose goal is to develop a plan for potable reuse (Miller, 2015). Similarly, water reuse regulations in the state of California have evolved as water availability for the growing population has become more challenging. Some of the most important IPR projects in the country are located in Los Angeles, Orange and San Diego counties. In 2014, regulations for groundwater recharge IPR projects were adopted, and by mandate of the state, the legislature is working on developing the country's first surface water augmentation regulations for IPR and investigating the feasibility of developing regulations for DPR by the end of 2016 (EPA, 2012). The state of Nevada has both guidelines and regulations, depending on the water reuse category, but they are limited to greywater reuse. The state of Hawaii has established only guidelines.

In the Pacific Northwest, Alaska does not have regulations that specifically address water reuse, and water reuse has generally not been implemented, whereas Idaho, Oregon and Washington have developed regulations on water reuse since the late 1980s, and water reuse has been a common practice since then (EPA, 2012).

At present, there are no federal regulations dictating criteria on DPR (EPA, 2012; NRC, 2012; Tchobanoglous et al., 2015; Tchobanoglous et al., 2011). Although it has not been officially recognized by state agencies, it is clear that de facto or unplanned potable reuse has historically accounted for significant potable water reuse. Because approximately 50% of the total wastewater effluent in the country receives secondary treatment (NRC, 2012), health risks are already present in at least half of the country's waters. The limited number of DPR projects is likely related to their perception as options of last resort for water utilities suffering from water scarcity. However, the small number of projects can also be related to a genuine lack of clear and standardized guidance on the design and operation of engineered natural systems and the low reliability of monitoring and control protocols prior to and after treatment to guarantee a 'fail-safe' condition of the system (NRC, 2012), particularly with regard to the effective removal of pathogens (Jin, Maleky, Kramer, & Ikehata, 2013; Miller, 2015). Regardless of the cause, the trend favours more consistent, planned DPR projects, as the costs of the projects pale in comparison to the cost of running out of water. By the end of 2016, California should be the first state with IPR regulations for surface water augmentation and will begin to develop DPR regulations, if found feasible, and it is expected that experienced states such as Florida, Texas and Arizona will follow the same trend. As water scarcity continues to challenge traditional reuse schemes, economic feasibility and technological reliability will be not only reachable but necessary.

Water ownership for reuse applications

In the US, each state determines ownership and use (or reuse) rights with respect to surface waters and groundwater within their boundaries. For surface waters, the riparian doctrine in the eastern US and the prior-appropriation doctrine in the western US determine water ownership. Most states address the ownership of wastewater for reuse projects under existing water rights systems, but a few have developed special water reuse regulations. As a general rule, ownership of water for reuse is often governed by the type of reuse, including IPR and DPR uses.

Riparian water rights states

Riparian law grants landowners whose property touches a natural watercourse the right to use a reasonable amount of water, and in times of shortage, each user's share is reduced according to reasonable-use criteria. This doctrine, followed in 29 states, generally east of the Mississippi River, has largely been replaced by a riparian permit system that requires landowners to obtain permits from a state agency in order to use water (Getches, 2009). Under riparian law, wastewater operators are generally able to retain rights to reuse water unless such reuse would cause unreasonable harm to downstream riparian rights holders. For example, Florida, which is a major water reuse state, has developed special rules encouraging water reuse. Florida cities are granted consumptive-use permits, including provisions for reusing treated wastewater, and they are able to retain ownership of wastewater for potable and nonpotable uses (FDEP, 2014).

Prior-appropriation water rights states

Twenty-one states in the western US have adopted prior-appropriation rules or have retained some riparian concepts in their water rights systems (Getches, 2009). Under prior appropriation, water is owned by the state and allocated to cities and other users based on a permit system. Often explained as 'first in time, first in right', these permits are prioritized based on seniority. Thus, when shortages occur, senior water rights users are protected and allocated their full amount of water, whereas junior water rights holders are prohibited from diverting or using water if such usage would harm senior rights holders.

Because of population growth over the last four decades, water demand in the western states has increased, stressing existing water resources, institutions and water laws. These new demands have provided the impetus for water reuse projects, thereby reducing the amount of water returned to rivers for downstream water users. An important question in prior-appropriation states is the extent to which water rights holders can retain and reuse water when such use would adversely affect other downstream users (Brockmann, 2011). Likewise, an important question is the extent to which downstream water rights can be guaranteed under future effluent conditions. Providing solutions to these challenges has been problematic for a water rights system that was established more than 100 years ago and was designed to accommodate agricultural and mining water use. State solutions to water ownership rights for reuse vary. Many western states have established, through court decisions or legislation, some balance between encouraging reclamation and reuse and protecting downstream water users. States favouring reuse tend to grant wastewater ownership rights to cities, meaning that they do not require the city to discharge water into a natural watercourse. Conversely, downstream water right holders cannot compel the city to discharge their wastewater back into the watercourse. Accounting for ownership differences, the following examples highlight several state responses to water reuse, including IPR and DPR.

Judicially determined ownership rule

Absent a permit requirement to return water to a stream, a water rights holder can generally reclaim water for direct potable or nonpotable reuse projects. The right of a city to retain ownership of treated wastewater for direct reuse has been judicially recognized in Arizona, Colorado, New Mexico and Montana (Arizona Supreme Court, 1989; Colorado Supreme Court, 1972; Montana Supreme Court, 1996; New Mexico Supreme Court, 1972). However, for indirect reuse projects, where wastewater is returned to a stream for later recapture, the general rule is that the water belongs to the state and can be allocated to other users.

Legislatively determined ownership rule

A number of western states statutorily clarify ownership of wastewater for reuse, including DPR and IPR. These statutes give cities the right to reclaim wastewater for reuse. California, for example, allows the owners of wastewater to retain control over the wastewater, but establishes procedures for evaluating impacts on downstream users before granting a reuse right (California Water Code §1210, 1211). Nevada requires cities to specify how the reclaimed water will be reused (Nevada Rev. Statutes §533, 440). The state of Oregon allows reuse, provided that fish and wildlife values are not adversely harmed (Oregon Revised Statutes §537.132). Utah requires a reuse permit to also conform to water quality requirements (Utah Code Ann. §73-3c-201, 202). The state of Washington requires the reclaimer to assess impacts on downstream users (Washington Revised Code §90.46.130).

Ownership rules for indirect reuse (potable or nonpotable)

Regardless of the judiciary and legislative cases mentioned above, most western states do not recognize retained ownership for indirect reuse (potable or nonpotable) once the water has entered a stream. Instead, they consider water returned to the watercourse or aquifer to be state water available for appropriation by others. However, Colorado and Texas have developed exceptions to this general rule. In Colorado, wastewater ownership is retained even when water is discharged into a river for later use or sale to other providers as 'developed water'. Water not native to the stream but imported from another basin is termed 'developed water', and it is available for indirect reuse (Colorado Supreme Court, 1996). Texas provides for greater indirect reuse of surface and groundwater by allowing a wastewater operator (e.g. a city) to retain ownership even if the water is discharged into a watercourse. Section 11.042 of the Texas Water Code allows a wastewater owner (city) to receive permission through a "bed and banks" permit to transport wastewater downstream for subsequent reuse or sale to others. Wastewater from groundwater, which is privately owned, can also be transported through a bed and banks permit for subsequent reuse or sale to others (§11.042 (b)).

In summary, state riparian and prior-appropriation water laws allow for direct potable reuse, but there are ownership barriers for indirect potable reuse. State water laws that address retaining of wastewater ownership for indirect potable reuse require clarification and possible expansion as IPR projects become more attractive and necessary, particularly in western states. The doctrines of water ownership will also need to be revised as a new paradigm of water management arises.

Case studies and lessons learned

Worldwide, IPR and DPR projects have recently been gaining momentum. Singapore's NEWater project provides ~60 MGD of reused water for industrial applications and a small amount for potable purposes (EPA, 2012). Namibia's Windhoek project revolutionized the water industry in 1967 as the first DPR project in the world (EPA, 2012). Israel has bolstered its agricultural production by optimizing its wastewater assets and today ranks among the highest users of reused water by capita and as a ratio to freshwater usage (Angelakis & Gikas, 2014). In Texas in 2014, the city of Wichita Falls fast-tracked an emergency 5 MGD DPR project as a result of the 2011 drought; this project has since transitioned to a 12–16 MGD IPR project (Steinle-Darling, 2015). Finally, the San Antonio Water System's 35 MGD direct nonpotable reuse system has allowed the city to conserve Edwards Aquifer groundwater and provide protection to several federally listed endangered species. The 130-mile recycled water pipeline has also helped secure major employers and further diversify the region's economy (SAWS, 2015).

For the purpose of celebrating the early adopters behind the projects and to highlight the importance of proactive and clear communication with a utility's customers throughout the process, this section will address specific case studies of firsts in potable reuse: the first planned IPR via groundwater recharge in California; the first planned IPR via surface augmentation in the US; the first planned IPR via surface augmentation in Europe; and the first sizeable DPR in the US.

Montebello Forebay Groundwater Recharge Project

Southern California's public discourse on water has largely been dominated by large importation projects carried out by the Metropolitan Water District of Southern California and the Los Angeles Department of Water and Power. However, three lesser-known water agencies in Southern California have been revolutionizing potable reuse for over 50 years, and their joint project has laid the groundwork for communities across California struggling in the current unprecedented water crisis (Johnson, 2010).

During the middle of the twentieth century, groundwater was being depleted across Southern California at an alarming rate. Groundwater production was twice the natural recharge, resulting in overdraft. Water levels declined up to 10 feet per year; wells went dry; and saltwater began to intrude. Beginning in the 1930s, various agencies began implementing solutions such as stormwater spreading grounds, saltwater barrier injection wells and court-ordered adjudications that set a cap on production (Johnson, 2009). This section focuses on an innovative potable reuse project that was conceived in the late 1950s, built in 1962 and became the first planned IPR project in California and one of the first in the US.

The Montebello Forebay Groundwater Recharge Project is a partnership between the Water Replenishment District of Southern California (WRD), which administers the overall management of the groundwater basin; the Los Angeles County Department of Public Works (DPW), which operates the system; and the Los Angeles County Sanitation Districts (CSD), which provide the recycled water. Water is treated three times at three different CSD water reclamation plants, discharged into a river channel or direct pipeline, and subsequently diverted and applied to one of two DPW spreading grounds, where it percolates into the groundwater basin (Gasca, Johnson, & Willardson, 2011). The Rio Hondo Spreading Ground is a 570-acre facility with 20 individual basins, and the San Gabriel Spreading Ground is a 128-acre facility with 3 individual basins (EPA, 2012).

Approximately 45 million gallons are applied to the spreading grounds daily, which accounts for ~35% of the total recharge to the groundwater basin. Both spreading grounds are operated under a wetting/drying cycle designed to optimize the inflow and discourage the development of vectors. Extensive monitoring is conducted at the water reclamation plants, at the headworks to the spreading grounds, and in the aquifers.

There is little evidence of public outreach between the time that the WRD was formed in 1959 and when Montebello Forebay was built in 1962, but the project has certainly set the standard for public outreach and education since its implementation. Thoughtful outreach has resulted in the steady increase in permitted recharge amounts. The WRD manages a Water Resources Committee that meets monthly, distributes both technical reports and informational brochures, and provides tours of the facilities.

Montebello Forebay ignited a movement in Southern California that immediately reversed a nearly 100-foot drop in aquifer levels (Johnson, 2009) and has since resulted in the construction of numerous groundwater recharge IPR projects. These projects have a combined offset of greater than 200 MGD, which otherwise would have been fulfilled by freshwater

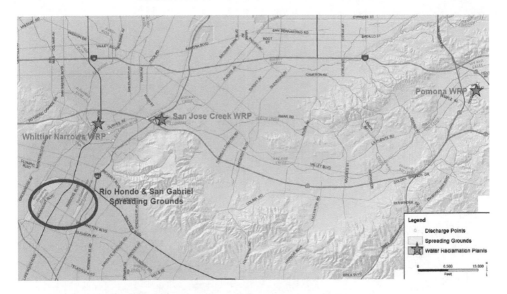

Figure 1. Map of the Montebello Forebay Groundwater Recharge Project (Gasca et al., 2011).

production/importation (Pecson, 2014). Additionally, during the 50 years Montebello Forebay has been in operation, numerous water quality and human health studies on the project have confirmed the safety of IPR via groundwater recharge. Figure 1 shows a map of the project.

Occoquan Policy

Located west of Washington, D.C., the Occoquan Reservoir is a valuable resource for the metropolitan area. The history of the management of this system is a shining example of regional collaboration. Rapid urbanization west of the nation's capital began in the 1960s and resulted in the degradation of the reservoir's water quality. Like other situations across the country prior to the Clean Water Act, the Occoquan Reservoir was being polluted by nominally treated effluents, in this case from 11 small wastewater treatment plants. In 1971, the Virginia Department of Environmental Quality (VDEQ) and the Virginia Department of Health (VDH) established the first planned IPR via surface augmentation in the US. The Occoquan Policy created the Upper Occoquan Service Authority (UOSA), a regional state entity mandated to collect and reclaim wastewater, and the Occoquan Watershed Monitoring Program (OWMP), which was tasked with monitoring water quality within the watershed and reservoir (EPA, 2012). Since the creation of these two state entities, the improvement of the water quality within the Occoquan system has benefitted Fairfax Water (which serves much of Northern Virginia) as well as the Potomac River and Chesapeake Bay further downstream.

UOSA's reclamation system (Figure 2) provides approximately 32 MGD of reliable and high-quality purified water on an annual average basis. However, the system has a capacity of 54 MGD and could be increased to accommodate an anticipated future demand of 65 MGD (EPA, 2012). During an average year, purified water discharged by UOSA accounts for approximately 10% of the water entering the Occoquan Reservoir. However, during drought years, the purified water discharged by UOSA has been documented to account for up to 80% of the water entering the Occoquan Reservoir (National Academy of Science, 2012).

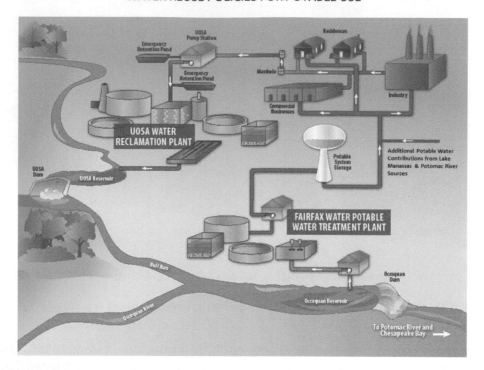

Figure 2. The Occoquan system in Virginia (EPA, 2012).

It is no coincidence that the first planned IPR in the US to utilize surface augmentation was a direct result of regional collaboration. The OWMP has a technical advisory panel that consists of representatives from the EPA, VDEQ, VDH, and Virginia Tech. Other regional issues related to the Occoquan system have been resolved via technical advisory groups, citizen action committees, and task forces.

When this project was first proposed, numerous public hearings were held to inform the public of what was being proposed and to give them a venue to voice their opinions. UOSA has provided tours of its facilities for more than 30 years, a commitment that has spanned generations and a testament to the successful implementation of the progressive Occoquan Policy.

Langford Recycling Scheme

Although the Occoquan Policy was largely implemented to improve water quality and foreshadowed the impending Clean Water Act, the Langford Recycling Scheme (Figure 3) was developed well after the water quality advances of the 1970s. Interestingly, the Langford Recycling Scheme was built to enhance supplies in a region not known for water shortages: the United Kingdom.

Essex & Suffolk Water (ESW) provides potable water to approximately 1.5 million people in the south-east of England (Essex & Suffolk Water, 2015). In response to drought, increasing demand (Hills, 2008) and a supply deficit in the 1990s (EPA, 2012), ESW pursued a scheme to reintroduce purified water upstream of its Water Treatment Works. Between 1993 and 1999, ESW conducted baseline studies, and in 2000 it was granted a permit from

Langford recycling scheme

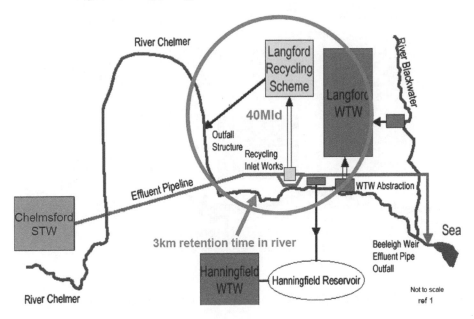

Figure 3. The Langford Recycling Scheme in south-east England (EPA, 2012).

the UK Environment Agency to discharge reclaimed wastewater. Construction began on the Langford Recycling Scheme plant, which became fully operational in 2003 (Lunn, 2013).

The plant recycles treated effluent from Anglian Water's Chelmsford Sewage Treatment Works by diverting it from a 15 km pipeline that discharges to the sea. Because the Sewage Treatment Works and pipeline were in existence in the 1990s, ESW had simply to build a pipeline and plant to divert and further treat the effluent. ESW discharges approximately 6 MGD (10.5 MGD capacity) of purified water into the River Chelmer, which is later diverted downstream to ESW's Water Treatment Works (EPA, 2012).

Notably, the Langford Recycling Scheme became the first IPR in Europe to utilize surface augmentation because it involved stakeholders in the process. ESW spent 10 years promoting the scheme to its customers and environmentalists and even built a mobile information workshop that visited all of the areas of its system that would be receiving the recycled water (Hills, 2008). There was some concern about drinking recycled water, as well as vocal concern about the possibility of siltation caused by reduced flows. However, ESW addressed those concerns and used the experience as a lesson for the promotion of future initiatives (Lunn, 2013).

Colorado River Municipal Water District, Big Spring, Texas

Situated in the Permian Basin of West Texas, the Colorado River Municipal Water District (CRMWD) serves approximately 350,000 people in the member cities of Big Spring, Snyder and Odessa, as well as the customer cities of Midland, San Angelo and Abilene (EPA, 2012). The CRMWD owns and operates three major surface water reservoirs on the Colorado River in Texas. Further downstream, Austin's Highland Lakes along the Colorado River serve as

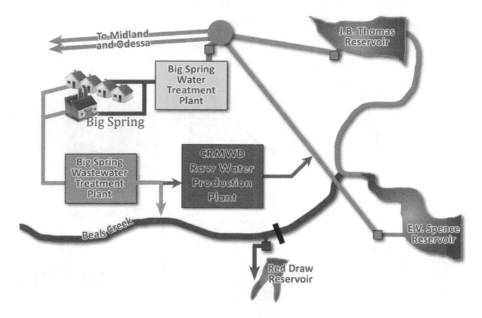

Figure 4. Schematic of Colorado River Municipal Water District's Reclaim Treatment Plant in Texas (EPA, 2012).

the city's sole source of potable water. However, Austin Water Utility's Water Reclamation Initiative supplies a modest 4 MGD of reclaimed water for nonpotable reuse, with plans for expansion (Austin Water, 2013). The CRMWD determined that most of the fresh groundwater in the Permian Basin had been developed; surface reservoirs were distant and at lower elevations than the population centres; and IPR was not desirable because the region experiences over 60 inches of evaporation per year (Martin, 2014). For those reasons, along with a prolonged drought since the 1990s (severe drought since 2011) and a rapidly increasing population associated with the oil and gas industry, the CRMWD set out to "reclaim 100 percent of the water, 100 percent of the time" (EPA, 2012).

Phase 1 of this ambitious plan involved diverting 2.5 MGD of treated effluent from the city of Big Spring's wastewater treatment plant, applying further treatment at the state-of-the-art CRMWD Reclaim Treatment Plant and finally blending that purified water with raw surface water in the CRMWD's existing raw water transmission pipeline. This water is then provided to its customer cities, where it is further treated to drinking water standards (Figure 4). Construction began in June 2011, and the project was fully operational by 2013 (EPA, 2012).

The CRMWD was able to build the first sizeable DPR in the US because of visionary leadership, proactive communication, and the inescapable reality of the region's water situation leading up to and during construction. In sharp contrast to the reactive and emergency nature of some DPR projects (Steinle-Darling, 2015), the CRMWD began planning for DPR in 2002 (Martin, 2014). West Texans' acute awareness of the value of water, CRMWD's transparency to its customers throughout the project, and CRMWD's proactive communication with regulators ensured the ultimate success of this pioneering project.

Potable reuse applications can be configured in seemingly countless different ways. While the Occoquan Policy was a regional solution to water quality concerns, the Langford Recycling Scheme was a strategy by ESW to enhance its water supplies. Although the

Montebello Forebay Groundwater Recharge Project uses spreading grounds, other IPR projects use injection wells. Whereas the Langford Recycling Scheme uses the banks of the River Chelmer to convey its purified water, the CRMWD uses its existing water transmission system. However, the success of all four projects has hinged on transparent and proactive communication throughout the life of the project.

Final remarks

Because water scarcity and environmental concerns challenge the traditional water supply paradigm, potable reuse, both indirect and direct, as a source of new or alternative potable water supplies has gained recent attention. Technology and engineering that guarantee the efficient removal of organics, chemicals and pathogens have already been developed and experience constant improvement. However, pending issues related to monitoring systems, standardized processing protocols, 'fail-safe' engineering systems, site-specific costs, and public acceptance are still under scrutiny. Likewise, because future alternative water sources for potable supplies will require more stringent water quality standards, the legal frameworks at the national and local levels need to be revised and adapted to the potential trend towards more 'direct' potable wastewater reuse. Traditional doctrines of water ownership (riparian and prior-appropriation rules) will require revision to allow the expansion of water reuse rights, particularly for IPR projects. Stakeholder engagement, public participation, and sustained communication are all key elements for a successful long-term water reuse strategy. Preparedness and policy development have proven to be more effective in the long term, compared to fast-track options designed to cope with a water shortage emergency with limited ability to follow a long-term plan. The present regulations and guidelines reviewed here seem to be directly proportional to the level of scarcity and water needs in particular states or regions. This fact may not necessarily represent a sustainable long-term or large-scale approach, as climate variability can always offer potential alternative scenarios of wetter periods in the future. However, the risk of relying on climate variability to avoid introducing the 'yuck factor' into the discussion only increases over time, and the increasing normality of water scarcity, population growth and climate change do not provide better scenarios for those who rely on traditional sources of water.

Disclosure statement

No potential conflict of interest was reported by the authors.

References

Angelakis, A. N., & Gikas, P. (2014). Water reuse: Overview of current practices and trends in the world with emphasis on EU states. *Water Utility Journal, 8*, 67–78.

Arizona Supreme Court (1989). Arizona Public Service Co.V.Long, 773 P.2d 988 C.F.R.

Asano, T., Burton, F. L., Leverenz, H. L., Tsuchihashi, R., & Tchobanoglous, G. (2007). *Water reuse: Issues, technologies, and applications*. New York, NY: McGraw-Hill.

Austin Water. (2013). Water reclamation initiative. Retrieved 2015, from https://www.austintexas.gov/department/water-reclamation

Brockmann, J. (2011). Water use and reuse: The new hydrologic cycle (Vol. 57): Rocky Mountain Mineral Law Institute.

Brooks, B. W., Riley, T. M., & Taylor, R. D. (2006). Water quality of effluent-dominated ecosystems: Ecotoxicological, hydrological, and management considerations. *Hydrobiologia, 556*, 365–379.

California Water Code §1210, 1211. Retrieved from http://ca.regstoday.com/law/wat/ca.regstoday.com/laws/wat/calaw-wat_DIVISION2_PART2_CHAPTER1.aspx#1.5

Colorado Supreme Court (1972). City and County of Denver v. Fulton Ditch Company, 506 P.2d 144 C.F.R.

Colorado Supreme Court (1996). City of Thornton v. Bijou Irrigation Co., 926 P2d 1. 65-78 C.F.R.

Cotruvo, J. A. (2015). Potable water reuse history and a new framework for decision making. *International Journal of Water Resources Development*. doi: 10.1080/07900627.2015.1099520

EPA. (2010). *LT2ESWTR toolbox guidance manual*. Cincinnati, OH: Environmental Protection Agency.

EPA. (2012). *2012 guidelines for water reuse*. (EPA/600/R-12/618). Washington, DC: U.S Agency for International Development.

EPA. (2015). Safe Drinking Water Act. Retrieved from http://water.epa.gov/lawsregs/rulesregs/sdwa/index.cfm#sdwafs)

Essex & Suffolk Water. (2015). At glance. Retrieved from https://www.eswater.co.uk/your-home/at-a-glance.aspx

FDEP. (2012). *2011 reuse inventory*. Tallahassee, FL: Florida Department of Environmental Protection.

FDEP. (2014). Florida's reuse activities. Retrieved from http://www.dep.state.fl.us/water/reuse/

Gasca, M., Johnson, T., & Willardson, B. (2011). *Keeping the water going: Challenges to recharge the montebello forebay*. Paper presented at the Managed Aquifer Recharge Symposium, Irvine, California.

Gerrity, D., Pecson, B., Trusell, R. S., & Trusell, R. R. (2013). Potable reuse trains throughout the world. *Water Supply: Research and Technology-AQUA, 62*(6), 321–338.

Getches, D. (2009). *Water law in a nutshell*. St. Paul, MN: West Pub. Co.

Gleick, P. (2003). Global freshwater resources: Soft-path solutions for the 21st century. *Science, 302*, 1524–1528.

Hills, S. (2008). Indirect Potable Reuse: UK Experiences & Lessons Learned. Paper presented at the WaterReuse Association/International Water Association.

Jin, Y., Maleky, N., Kramer, N. A., & Ikehata, K. (2013). Health effects associated with wastewater treatment, reuse, and disposal. *Water Environment Research, 85*, 1954–1977.

Johnson, T. (2009). Recycled water for groundwater recharge: Innovative recharge projects and source water implications (W. R. D. o. S. California, Trans.) (pp. 1–39). Los Angeles, CA: California Water Resources Department.

Johnson, T. (2010). Recycled Water for Recharge: A Growing Resource for Sustainable Groundwater Replenishment: Water Replenishment District of Southern California.

Kenny, J., Barber, N., Hutson, S., Linsey, K., Lovelace, J., & Maupin, M. (2009). *Estimated use of water in the United States in 2005*. Reston, VA: USGS.

Lunn, M. (2013). *Langford recycling scheme*. UK: Northumbrian Water Group (NWG).

Martin, L. (2014). Texas leads the way with first direct potable reuse facilities In U.S. Water Online. Retrieved from http://www.wateronline.com/doc/texas-leads-the-way-with-first-direct-potable-reuse-facilities-in-u-s-0001

Meehan, K., Ormerod, K. J., & Moore, S. A. (2013). Remaking waste as water: The governance of recycled effluent for potable water supply. *Water Alternatives, 6*, 67–85.

Miller, W. (2015). Direct potable reuse: Its time has come. *Journal - American Water Works Association, 107*(5), 14–20.

Montana Supreme Court (1996). In Re City of Deer Lodge, Case B-No 97514-76G C.F.R.

Nevada Rev Statutes §533,440. Retrieved from http://law.justia.com/codes/nevada/2013/chapter-533/statute-533.440

New Mexico Supreme Court (1972). Reynolds v. City of Roswell, 654 P.2d 539 C.F.R.

NRC. (1998). *Issues in potable reuse*. Washington, DC: The National Academies Press.

NRC. (2012). *Water reuse: Potential for expanding the nation's water supply through reuse of municipal wastewater*. Washington, DC: The National Academies Press.

NWRI. (2012). *Ultraviolet disinfection guidelines for drinking water and water reuse*, 3rd edn. Fountain Valley, CA: National Water Research Institute.

Oregon Revised Statutes §537.132. Retrieved from https://www.oregonlegislature.gov/bills_laws/lawsstatutes/2013ors537.html

Ormerod, K. J., & Scott, C. (2012). Drinking wastewater: Public trust in potable reuse. *Science, Technology, & Human Values, 38*, 351–373.

Pecson, B. (2014). *Potable reuse in California: Lessons learned and the path forward water innovation summit* (pp. 1–47). Albuquerque, NM: Trussell Technologies Inc..

Rock, C., McLain, J. E., & Gerrity, D. (2012). Water Recycling FAQ (College of Agriculture and Life Sciences, Trans.) (pp. 1–8). Tucson, AZ: Arizona Cooperative Extension.

SAWS. (2015). Water Recycling. Retrieved from http://www.saws.org/Your_Water/Recycling/

Schroeder, E., Tchobanoglous, G., Leverenz, H. L., & Asano, T. (2012). *Direct potable reuse: Benefits for public water supplies, agriculture, the environment, and energy conservation* (p. 15). Fountain Valley, California: National Water Research Institute.

Steinle-Darling, E. (2015). The many face of DPR in Texas. American Water Resources Association, March, 16-20.

Tchobanoglous, G., Cotruvo, J., Crook, J., McDonald, E., Olivieri, A., Salveson, A., & Trusell, S. (2015). *Framework for direct potable reuse*. Alexandria, VA: WateReuse Research Foundation.

Tchobanoglous, G., Leverenz, H. L., Nellor, M. H., & Crook, J. (2011). *Direct potable reuse: The path forward*. Washington, DC: WateReuse Research Foundation and Water Reuse California.

Texas Water Code §11.042 (b). Retrieved from http://codes.lp.findlaw.com/txstatutes/WA/2/B/11/B/11.042

Upper Occoquan Service Authority. (2014). Centerville, VA: Upper Occoquan Service Authority. Retrieved April 10, 2015, from http://uosa.org

Vigneswaran, S., & Sundaravadivel, M. (2004). Recycle and reuse of domestic wastewater. In S. V. Vigneswaran (Ed.), *Encyclopedia of life support systems*. Oxford, UK: EOLSS.

Washington Revised Code §90.46.130. Retrieved from http://app.leg.wa.gov/rcw/default.aspx?cite=90.46.130

WateReuse Association. (2011). Sustainable solutions for a thirsty plant 2015, from http://www.wateruse.org/information-resources/about-water-reuse/faqs-0

Western Resource Advocates. (2012). Meeting future urban water needs in the Arkansas River Basin. Retrieved April 20, 2015, from http://westernresourceadvocated.org/water/fillingthegap/FTG-ARK-Reuse.pdf

Utah Code Ann. §73-3c-201 and 202. Retrieved from http://le.utah.gov/xcode/Title73/Chapter3C/73-3c-S201.html

Common or independent? The debate over regulations and standards for water reuse in Europe

John Fawell, Kristell Le Corre and Paul Jeffrey

Cranfield Water Science Institute, Cranfield University, UK

ABSTRACT

Although unplanned water reuse has been practised across Europe for decades, multiple stresses on water supply and demand over recent years have led to the development of many planned reuse schemes. Despite this development, the legislative and regulatory regimes required to underpin a growing water reuse sector have arguably failed to emerge. The reasons for this and the cases for and against pan-European water reuse regulations are explored and debated. The conclusions highlight several challenges for politicians and policy makers if appropriate regulatory systems and water quality standards are to be provided which support the embryonic European water reuse sector.

Introduction

In the 28 member states of the European Union (EU), approximately 81% of the freshwater used for energy production, agriculture, public water supply and industry originates in surface water bodies, with groundwater being the primary source for public water supply (European Environment Agency [EEA], 2009). Abstraction pressures on surface water bodies are likely to increase in the medium-to-long term as a result of population growth and the impact of climate change (Alcalde Sanz & Gawlik, 2014). As well as being sources of potable and irrigation water, rivers are widely used as receiving-water bodies for treated wastewater effluent. In effect this means that unplanned indirect reuse has been practised for many decades where river water used for water supply is abstracted downstream of cities discharging their treated wastewater into the river (Asano, 1998). The increasing pressures on water sources and the need to dispose of treated wastewater effluent have driven significant improvements in wastewater treatment as well as drinking water treatment, although these have not been specifically driven by the potential for planned wastewater reuse. Wider implementation of water reuse would have distinctive benefits for the European environment and economy by providing additional water resources at competitive cost and reducing the demand on limited freshwater resources. For instance, Hochstrat, Wintgens, and Melin (2008) estimated that water savings resulting from the implementation of water and wastewater reclamation in Europe could reach up to 1.5% by 2025, with specific southern European countries such as Malta, Spain and Cyprus having the potential to reduce by up to 17%, 7% and 3%, respectively, their water abstraction as a result of wider use of water reclamation and reuse.

Although Europe has a number of areas where water reuse is practised (either directly or indirectly), the lack of any Europe-wide standards or guidelines for reuse, for either potable or agriculture applications, is viewed as one of the major barriers to the development of the water reuse sector (Technopolis Group, 2013). As an agglomeration of 28 member states, many of which have significant internal variation in terms of culture and history, Europe (in the form of the EU) provides a particularly delicate patchwork of norms and behaviours within which to develop a consistent and coherent water reuse framework. Any standards or water quality guidelines developed through such a framework will need not only to cover a wide range of water reuse purposes, from agricultural to urban, but also to be both acceptable and equitable across all 28 member states. Failure to achieve this may for example compromise the free trade of agricultural products irrigated with treated wastewater from countries applying less stringent health and safety standards than those applied in the countries importing the products (European Commission, 2014). In addition, the EU has a number of transboundary water bodies and rivers, such has the Danube River, which flows through seven EU countries, which further complicate the position. EU-level regulatory changes are often slow to be implemented, as they (quite rightly) require extensive review, revision, and impact assessment. The fact that the EU is such a large trading bloc also impacts other countries in the region that are not EU members. The importance of ensuring that regulation facilitates rather than burdens reuse schemes was starkly illustrated in a recent report commissioned by the EU itself, pointing out that in three large EU countries (France, Italy and Greece) overly stringent non-potable reuse quality standards were seen as a major barrier to the further development of reuse projects (BIO by Deloitte, 2015a). Variations in these national standards are briefly discussed in the following section.

The EU currently relies on several major pieces of legislation to regulate the natural and engineered water cycles. Member states are expected to adopt these directives and incorporate their requirements into national legislation, with the European Commission itself overseeing their implementation. Two of these are of particular significance with respect to water reuse. Council Directive 2000/60/EC (the Water Framework Directive) (EC, 2000) establishes a framework for action in the field of water policy and indirectly recognizes reuse as a strategy for increasing water availability, which thereby contributes to the good quality status of water bodies. It also refers (in Annex VI:x, p. L327/64) to "efficiency and reuse measures". A second significant influence is Council Directive 91/271/EEC (the Urban Waste Water Treatment Directive) (EEC, 1991), which relates to wastewater treatment and discharge. Article 12 (p. L135/42) of this directive states that "treated wastewater shall be reused whenever appropriate", implying that wastewater reuse is acceptable inasmuch as it does not breach other EU legislation or national laws. Other relevant EU ordinances include the Drinking Water Directive (80/778/EC, revised with 98/83/EC)(EC, 1998), which sets out the quality of water intended for human consumption, and the Groundwater Directive (2006/118/EC) (EC, 2006), which seeks to protect groundwater against pollution and deterioration.

Of particular significance to reuse schemes which rely on the use of an environmental buffer (e.g. a river or aquifer), the anticipated revision of the Priority Substances Directive

(Directive 2013/39/EU, a so-called 'daughter' directive of the Water Framework Directive) (EC, 2013) will impose water quality standards for emerging pollutants that may influence indirect potable reuse scheme design, and treatment technology selection in particular. The impact of this legislation will depend on the type of water body being used as the environmental buffer and the details of scheme design. Indeed, the Priority Substances Directive could well strongly influence the economics and therefore the attractiveness of large-scale reuse schemes across the continent.

Importantly, none of the statutory instruments listed above are overtly directed at regulating or promoting wastewater reuse as such. The EU has historically shied away from intervening in the fledgling European water reuse sector, preferring that national administrations take the lead in setting appropriate laws and regulations. It has not imposed universal non-potable water quality criteria, nor has it provided the sort of enabling legislation and guidance which might encourage reuse at large or small scales.

This is not to say that the EU's various functional and policy bodies have ignored water reuse. The European Commission, which is the executive arm of the EU, has funded research and innovation activities to promote the development of reuse strategies and advanced treatment technologies and has developed appropriate risk management strategies for reuse schemes, e.g. the Aquarec (Framework V / EVK1-CT-2002-00130), Reclaim Water (Framework VI / #18309), and Demoware (Framework VII / #619040) projects. It has also encouraged and responded positively to the inclusion of reuse within wider analyses of water policy and catchment-based management strategies. The most recent of these initiatives involves the widely referenced Water Blueprint (European Commission, 2013), which makes it clear that reuse of wastewater should be a major consideration in improving water efficiency in the EU and recommends that reuse be particularly focused on irrigation and industrial uses. It goes on to note that the limited extent of such schemes in the EU appears to be due to a lack of common EU environmental and health standards for reused water and the potential obstacles to the free movement of agricultural products irrigated with reused water. Several initiatives have been catalyzed by the Water Blueprint reports. The European Commission's Directorate-General for the Environment conducted a public consultation in late 2014, which concluded that the principle of reusing wastewater has attracted widespread acceptance within the EU despite concerns regarding its use for food crop irrigation and drinking water. In addition, there was a substantial majority of opinion that considered regulation essential to promoting reuse in the EU (BIO by Deloitte, 2015b).

The European Commission is also working with the World Health Organization (WHO), which is considering revisions to its Guidelines for the Safe Use of Wastewater, Excreta and Greywater (WHO, 2006). In addition, the commission has established a close collaboration with the European Innovation Partnership on Water, which is a platform to facilitate innovation in all parts of the European water industry. In particular, collaboration with two of the European Innovation Partnership on Water action groups – on industrial water reuse and recycling, and on water and irrigated agriculture – is delivering welcome support to the industrial and agricultural sectors. However, it is also worth noting that, despite the WHO (2006) standards, point-of-use water quality criteria across different regions of the world are variable. WHO is currently reviewing its standards for reuse in agriculture and for other purposes, and the outputs of this activity are likely to have a significant impact on European attitudes.

Variable point-of-use non-potable water quality standards across the EU

Untreated wastewater from municipal or industrial origins contains a range of hazards in the form of pathogens such as viruses and bacteria, and chemicals such as pharmaceutical substances, hormones and heavy metals. Wastewater reclamation, if not appropriately managed, could therefore present a threat to public health and the environment (Salgot, Vergés, & Angelakis, 2003). The potential for both microbial and chemical contaminants to have adverse effects on human health will depend on transport and routes to human contact as well as on levels of exposure. Possible routes of exposure to pathogens and/or compounds of concern may include ingestion through the consumption of drinking water or crops, meat, or fish produced using reclaimed wastewater; skin contact or inhalation through recreational activities, e.g. irrigation of public parks, sports fields and golf courses; or direct contact through professional activities, e.g. in agricultural workers (Godfree & Godfrey, 2008).

In the EU, a number of countries, including Cyprus, France, Greece, Italy, Portugal and Spain, have developed point-of-use standards for non-potable water applications in their jurisdiction (Paranychianakis, Salgot, Snyder, & Angelakis, 2015). Among the requirements for the monitoring of treated wastewater to be reused, microbiological parameters are mandatory in all countries. The indicator typically used to evaluate the microbial quality of the reclaimed water is *E. coli,* which is considered more accurate than total coliforms or faecal coliforms in describing the microbial contamination of waters (Alcalde Sanz & Gawlik, 2014). To illustrate, the limit values for *E. coli* for unrestricted irrigation vary between ≤ 5 cfu/100 mL (in 80% of samples) in Greece to ≤ 250 cfu/100 mL per week in France (Table 1). Paranychianakis et al. (2015) indicate that some of these countries, such as France, include additional parameters, such as the irrigation methods used and type of irrigated crops, soil properties and sampling frequency, etc., to better prevent risks. In some cases additional water quality parameters are specified that can add significantly to the costs of monitoring. For example, France defines a total of 6 water quality parameters, while Italy, Greece and Spain each include over 50 (Table 1). However, France also requires that the sludge from the wastewater treatment works be monitored. In most cases there are additional stipulations, such as water application control measures, ensuring that there are no cross-connections, and limits on the type of irrigation method (e.g. banning spray irrigation).

While the EU has introduced a number of binding policies in relation to the water environment, including the management of wastewater, individual member states are responsible for implementation at the national level, with consequent variation (to a wider or lesser degree) across the 28 member states. Differences in attitude to specific aspects of EU-level policies also result in variations in the tone and emphasis of national implementation legislation and regulation. Where there exists a policy vacuum at European level (as is the case with water reuse standards), these differences can become both evident and significant.

As noted above, a recognized challenge for the EU is for the 28 member states to agree on a common set of non-potable water quality criteria that can transcend these differing attitudes, which often reflect differing priorities, legacies and capacities in member states. The fact that several states have already established their own non-potable water quality criteria makes this process more difficult, as there may be limited willingness in those states to compromise and adapt their existing standards to a unified EU version. There are also differences in the attitudes of consumers between different countries (often between the north of Europe and the south – see Nocella, Boecker, Hubbard, & Scarpa,

Table 1. Microbiological standards for unrestricted irrigation in Cyprus, France, Greece, Italy, Portugal and Spain. Adapted from Paranychianakis et al. (2015).

	Cyprus[a]	France	Greece	Italy	Portugal	Spain[c]
Microbiological indicators						
E. coli (cfu/100 mL)	–	≤250	≤5 (for agricultural crops)	≤10	–	≤100
Total coliforms (cfu/100 mL)	–	–	2 (for urban irrigation)	–	–	–
Faecal coliforms (cfu/100 mL)	≤5	–	–	–	≤100	–
Helminth eggs (eggs/L)	0	–	–	–	≤1	<0.1 (0 for urban uses)
Additional information						
Frequency of analysis[b]	1/15 days	1/week	1–4/week depending on population served	Frequency not considered – to be determined by facility managers	Frequency not considered – to be determined by facility managers	1–3/week
Additional parameters included in the standards	–	–	64 (incl. heavy metals, metalloids, toxic substances incl. priority substances)	53 (incl. heavy metals, metalloids, toxic substances incl. priority substances)	Heavy metals, metalloids, toxic substances incl. priority substances[b]	Up to 60 (incl. heavy metals, metalloids, toxic substances incl. priority substances)
Additional requirements		Distance between irrigated areas; slope of irrigated fields			Type of soil (irrigation not allowed in karstic geological formations); slope of irrigated fields (not authorized if slope > 20%)	Monitoring at the point of use

[a]Irrigation forbidden of leaf vegetables and other vegetables eaten raw.
[b]Source: Alcalde, Sanz, and Gawlik (2014).
[c]Term 'unrestricted irrigation' not described for Spanish criteria.

2012) and in abilities to police and enforce regulations. Consequently, there needs to be a compromise between excessive precaution and insufficient safety in developing regulations. Where there is a widely agreed set of international norms this can be much more easily overcome, but the WHO guidelines, which are what Europe usually looks to, have proven to be difficult to implement, and they do not cover many emerging chemical contaminants of concern. The process of revision of the WHO guidelines is unlikely to be complete for some time.

Despite the lack of specific EU reuse regulations, wastewater reuse schemes have been successfully implemented under country-specific or even regional guidance (Hochstrat, Wintgens, Melin, & Jeffrey, 2006). Among these are the Torreele/Wulpen (Belgium) aquifer recharge scheme for indirect potable reuse, the aquifer recharge site of Tossa de Mar (Spain) for urban reuse, and the wastewater recycling for agricultural irrigation schemes in Milan, Italy, and Braunschweig, Germany (Table 2).

As can be seen in Table 2, a lack of EU-level regulation has arguably not stopped the development and implementation of reuse schemes, but is it hindering the development of a more confident and effective water reuse sector? Is it slowing down the development

Table 2. Examples of wastewater reclamation schemes in Europe.

Location	Year	Owner	Drivers for implementation	Reuse purpose	Source	Treatment	Volume of treated wastewater and/or reclaimed water	Costs (€)	Benefits	Reference
Amathus-Limassol, Cyprus	1992–2013[a]	Sewerage board of Limassol (SBLA)	Severe drought leading to water restrictions	Non-potable (urban and agricultural reuse)	Municipal wastewater	Primary treatment	Capacity of the Limassol WWTP in 2008: 40,000 m³/day (270,000 p.e.)[c]	Capex (1995) for tertiary treatment plant in 1995[g]: CYP 1.85M(approx. €3.16M in 2008)[b]	Freshwater resources savings Preservation of the natural environment	Papaiacovou & Papatheodoulou (2013)
			Increasing water demand (population growth and tourism)			Secondary treatment (conventional activated sludge)	Reclaimed water production: 6.7 MCM/y in 2011[c] (4.8 MCM used for agricultural purposes, 1.2 MCM discharged to sea, 0.7 MCM sent to Polemidia dam)	Customers are charged for the use of recla med water; in 2008, the price varied from 0.05 €/m³ for agricultural irrigation to 0.21 €/m³ for golf course irrigation[c]	Conservation of the sustainable development of the region	
Wulpen-Torrele-St André, Belgium	2002	IWVA	Aquifer depletion and pollution (saline intrusion)	Potable(indirect – groundwater augmentation)	Municipal wastewater	Tertiary treatment (sand/gravel filtration + disinfection)			Improvement of living standards of the region's inhabitants	Van Houtte, Cauwenberghs, Weemas, & Thoeye, (2012)
			Increasing water demand (population growth and tourism)			Tertiary treatment (disinfection; membrane filtration [UF/RO])	WWTP capacity: 2.3 MCM/y	Capex: €7M	35–40% of IWVA's annual drinking water production achieved by the scheme	
			Risks of saline intrusion in aquifer			Infiltration to aquifer via pond	Reclaimed water production: 1.8 MCM/y in 2011[d]	Cost to produce and infiltrate water: 0.64 €/m³ (2011)[d]	Improvement in the ecological value of the aquifer recharge site	
						After abstraction, aeration, rapid sand filtration, storage, and UV disinfection prior to distribution		The cost for recycled water is recovered from the drinking water price	Improvement of drinking water quality	
								A 21% increase in drinking water price after the scheme implementation was reported[e] In 2011, the price was 1.75 €/m³, compared to an average of 1.83 €/m³ in the Flanders region[d]		

(Continued)

Table 2. (*Continued*).

Location	Year	Owner	Drivers for implementation	Reuse purpose	Source	Treatment	Volume of treated wastewater and/or reclaimed water	Costs (€)	Benefits	Reference
Braunschweig, Germany	1979–2005[a]	Stadtentwässerung Braunschweig (operator: Abwasserverband Braunschweig, n.d.)	Need for water and fertilizers in an area where soils are poor; Environmental protection and health risks management (odour, heavy metals, groundwater pollution)	Non-potable (agricultural reuse)	Municipal wastewater	Primary treatment; Secondary treatment (conventional activated sludge with nitrification and denitrification stages)	Braunschweig WWTP capacity in 2008: 22 MCM/y (385,000 p.e.) 2/3 of the treated wastewater is used to irrigate 3000 ha of agricultural lands; 1/3 is further purified by infiltration fields before discharge or further use	Data not available	Irrigation of lands with reclaimed water having fertilizing properties; Prevention of contamination of surface water bodies	Ternes, Bonerz, Herrmann, Teiser, & Andersen (2007); Abwasserverband Braunschweig, n.d.
Clermont-Ferrand, France	1996	Clermont-Ferrand Municipality	Pollution of the river due to the discharge of wastewater effluent; Reduction in river flow and degradation of its quality due to agricultural irrigation; Regular drought events	Non-potable (agricultural reuse)	Municipal wastewater	Primary treatment; Secondary treatment (conventional activated sludge); Lagooning (surface 13 ha, 312,000 m³)	Clermont WWTP capacity: 64 MCM/y (425,000 p.e.) Volume of water used for irrigation: 1.1 MCM/y on average between	Capex (1996): €30M (WWTP) and €5.3M for the reclamation scheme Opex (1996): €86K + 0.050 €/m³ as energy	Distribution of water for irrigation to 50 local farmers; Moderate investments and operational costs; Preservation of the environment	Loubier & Declercq (2014)
Milan–San Rocco, Italy	2004	Suez Environnement–Degremont Italia; Consortium di bonifica Est Ticino Villoresi (distribution)	Pollution (discharge of raw wastewater into the environment); Contamination of water used for agriculture; Recurrent periods of drought; Increasing pressure on groundwater resources	Non-potable(agricultural reuse)	Municipal wastewater	Primary treatment; Secondary treatment (conventional activated sludge with nitrification and denitrification stages); Tertiary treatment (rapid sand filtration + UV disinfection)	San Rocco WWTP treatment capacity from 350,000 m³/day to 1.04 MCM/day Average production of tertiary effluent between 2005 and 2010: 96 MCM/y	Capex:€184M (WWTP)[f] Opex (2009): €567K for maintenance, €900K for energy, €157K for natural gas, €75K for chemicals[f]	Improvement of the chemical & microbiological quality of surface water bodies; Restoration of the biodiversity of surface water bodies; Supply of high-quality water for agricultural irrigation at no cost to farmers	Mazzini, Pedrazzi, & Lazarova (2013)

Table 2. (*Continued*).

Location	Year	Owner	Drivers for implementation	Reuse purpose	Source	Treatment	Volume of treated wastewater and/or reclaimed water	Costs (€)	Benefits	Reference
Tossa de Mar, Spain	2003	Consorci Costa Brava	Over-exploitation of the Tordera River's aquifer for drinking water supply as a result of tourism	Non-potable (urban reuse)	Municipal wastewater	Tertiary treatment (coagulation/flocculation + rapid sand filtration; disinfection [sodium hypochlorite and UV])	Average volume of reclaimed water produced in 2009: 80,000 m³	Capex (2009): €837,000 for the water reclamation plant and recycled water distribution network	Reduction in freshwater consumption through the use of reclaimed water for various activities (municipal services, landscape irrigation, etc.)	Sala (2010)
			Severe droughts in the late 1990s and early 2000s						Environmental protection (restoration of the Sa Riera park and local stream)	Mujeriego, Sala, & Couso, (2011)
			Reduction of groundwater levels and groundwater quality deterioration							
Old Ford Water–London Olympic Park, UK	2011	Thames water utilities	Droughts	Non-potable (urban reuse)	Raw wastewater	Pretreatment stage (septic tanks)	Reclamation plant capacity: 574 m³/day	Capex (2012): £7M (approx. €8.6 based on average 2012 £-to-€ exchange rate)	In 2014, the scheme recycled 41,000 m³ of treated wastewater to irrigate the parklands and 4200 m³ for non-potable use at the Copper Box sport venue (equivalent to 19% of the site's water consumption), with a 40% water reduction in potable water use	Knight, Maybank, Hannan, King, & Rigley (2012)
			Rapid population growth			Biological treatment (membrane bioreactor)				
						Polishing using granulated activated carbon filtration and disinfection (chlorination)				

Notes: WWTP: wastewater treatment plant; p.e.: population equivalent
[a]Construction/implementation in phases.
[b]Fixed exchange rate of CYP 0.585 to EUR 1.00 on 1 January 2008 (date of Cyprus' entry into euro zone).
[c]Larcou Yiannakou (2012).
[d]Van Houtte & Verbauwhede (2012).
[e]Otoo, Mateo-Sagasta & Madurangi (2015).
[f]Casiraghi (2014).
[g]European Commission (1999).

of more progressive water policies? And why not leave it up to individual European nation-states to set their own standards for recycled water?

What makes for effective regulation of water reuse schemes?

Regulation has a major influence on the feasibility, implementation and operation of water reuse schemes. A clear definition of realistic standards to protect the environment and human health, and guidelines for the credible operation and monitoring of schemes, provide public and commercial stakeholders with the confidence needed for investment. If licensing is to be the primary tool for regulatory control, then details of the expected spread of risk and responsibilities are required. However, regulatory policy needs to be flexible and robust to reflect the variability of scheme context. The establishment of clear standards for the quality of water provided for non-potable uses is an important prerequisite for a workable water reuse sector. Such standards provide both an operational performance target for scheme developers and confidence for water users. The legal position of companies offering non-potable water services is severely compromised in the absence of clear and binding regulations which aim to protect public health and ensure the safe operation of reuse schemes. In this context it is perhaps unsurprising that those countries with benchmark water reuse operations (e.g. the US, Australia, Cyprus) have strong and well-established non-potable quality criteria and mature governance arrangements. Mature water reuse sectors which make substantial contributions to resource management do not operate in a regulatory vacuum. For example, there are long histories of reuse regulations in many US states, such as Texas and California (Sanchez-Flores, Conner, & Kaiser, 2016). In addition to the underpinning regulations around process and responsibilities, rules and regulations relating to consumption and supply are also often very comprehensive.

An effective regulatory regime for water reuse schemes at the EU level which provides common targets for water quality and risk management is desirable for three main reasons. First, the provision of a firm legal basis to protect the health of the public and the environment is vital for any sector involved in the management and anthropogenic use of natural resources. The absence of such a robust legal architecture within which commercial and public bodies can operate erodes the confidence and conviction needed by potential reuse scheme developers; uncertainty reigns, and there are no independently set performance objectives upon which to found risk management practices.

This link between regulation and risk management is central to the effective operation of engineered systems which deliver utility for citizens on a large scale through the management of natural resources. Although several EU member states have legislated requirements for risk-based approaches to drinking water supply (e.g. the UK, Netherlands, Norway and Estonia), there is no EU-wide obligation, and there has been extensive debate about what an appropriate risk assessment method for water reuse schemes might look like (Nandha, Jeffrey, Berry, & Jefferson, 2013), particularly within a wider context of drinking water safety plans (Goodwin, Raffin, Jeffrey, & Smith, 2015). Although the relative suitability of different risk frameworks, such as 'hazard and operability study' and 'hazard analysis and critical control points', has been explored in relation to water reuse schemes, there is some consensus emerging around the adoption of modified water safety plans. Water safety plans provide a holistic approach to water service risk management by determining whether the water supply chain as a whole can deliver water of sufficient quality, ensuring the effective monitoring of those control measures in the supply chain that are of particular importance in securing water safety, and setting out management plans describing the actions to be undertaken, from normal conditions to extreme events (Bartram et al., 2009). Originally developed as a multiple-barrier risk management approach (i.e. from source to tap) for drinking water treatment plants to protect public health (Almeida, Vieira, & Smeets, 2014), the adaptation of water safety plan principles to broader applications, and more specifically water reuse, that involve additional water management challenges, such as public safety and environmental protection, has attracted much attention in recent years (Goodwin et al., 2015). For example, in 2008, the Queensland (Australia) government reinforced its legislation for the protection of public health by introducing new measures in its Water Supply (Safety and Reliability) Act 2008 concerning the use of recycled water which may end up in the drinking water supply chain (Roux et al., 2010). This implied the development of water management plans as a prerequisite to the approval of water recycling schemes. These plans, which include a risk-based approach for the management of recycling schemes (Roux et al., 2010), were derived from the Australian guidelines for water recycling (NRMMC, 2008), which, like water safety plans, are based on hazard analysis and critical control points.

Second, effective regulation moderates the perverse or conflicting incentives which can appear in sectors and markets and which lead to undesirable social or economic outcomes (as explained in Bakker, 2003). The provision of appropriate incentives for companies delivering services is a central tenet of regulation theory, a point succinctly made by the WWF when commenting on the role of the financial regulator in the UK. Stating that "companies must be given better incentives to manage water resources sustainably" and urging the removal of perverse incentives and the provision of rewards for "companies that invest in creative and innovative ways to reduce their impact on the environment", they crystalize the relationships between regulation, innovation and stewardship of the natural environment (WWF, 2010). However, in order for regulation to play this role, it must both understand the impact of regulatory interventions and anticipate institutional responses. This is a non-trivial challenge for those charged with developing and implementing regulatory regimes, and made even more difficult in the case of water reuse by the fact that the activities being regulated are often novel and have only sparse precedent. Under such circumstances governance bodies are perhaps understandably cautious and conservative.

The third principal argument for the development of an effective regulatory regime for water reuse schemes at the EU level is that geographical heterogeneity in regulation would have unwelcome consequences for European business and communities. For example, variability in non-potable water quality standards for agricultural use across the continent has the potential to damage the free movement of goods across Europe's internal borders as consumers in one part of Europe become anxious about perceived lower standards in other parts of the continent. This issue is potentially damaging to the single-market principle held so dear by the EU and is perhaps the reason why there has been more interest in developing water quality standards for agricultural water reuse than for the non-potable and potable municipal sectors.

On the other hand, of course, there are a number of well-understood disadvantages to a pan-European regulatory approach to water reuse which are worth articulating. We are entering a period in which there a number of uncertainties regarding some aspects of the science surrounding possible health risks from reuse of wastewater. In Europe concern over chemical contaminants, including emerging contaminants, remains a perceived problem, and data confirming or refuting whether or not this is really a problem and under what circumstances are required. Such a situation creates difficulties in developing regulations, and there is a danger that some member states who have less pressure on water resources may seek to propose a very precautionary approach. Such a situation would create significant tensions between member states with differing pressures on water resources.

There would also be difficulties for member states with existing standards, because these would normally be superseded by EU regulation. For those users who have invested in treatment and monitoring to achieve standards in their country that are higher than the final EU-wide standards, there is an issue of wasted investment. For others, the contrary may be true, and they may have to add further investment in treatment and irrigation systems that might render existing investment redundant. While such situations are not uncommon in European negotiations, they do make the negotiations more difficult. In addition, some users who are able to operate satisfactorily in their own country and are content with the local market may find that they are subject to increased monitoring and verification requirements that will simply add cost to their operations.

In addition, there may be issues between different member states due to the variability of the source water quality for reuse that could result in calls for significant changes to treatment and control of inputs. While this may be desirable to provide reassurance and would have additional benefits in improving the quality of receiving waters for the streams that are not going for reuse, it requires a very long-term approach, with substantial requirements for investment. Such calls could be a disincentive for the introduction of reuse in the short term.

Finally, the introduction of EU-wide regulation could be costly in terms of administration and monitoring and could also be unhelpful in the development of other water-conserving options by diverting attention and resources. This risk is of course a recipe for paralysis, as fear of compromising parallel or alternative strategies incentivizes procrastination, indecisiveness and inaction. European politicians are well able to develop clear policy objectives within complex contexts and to develop instruments to pursue realization of those objectives. The prioritization of incentives and regulations to shape desirable responses to Europe's degenerating water balance is the policy challenge. Trading off the first- and second-order impacts of preferred incentives against lost opportunities and unavailability of resources in other areas will expose the wider costs of candidate instruments.

Conclusion

The foregoing reflections on potential European approaches to regulating water reuse schemes are informed by a growing need for action. As the impacts of climate change and population growth and relocation transform the geography and temporality of Europe's supply–demand balance, water service providers are looking for new ways to enhance resource availability. Treating water to the quality needed for specific applications (and thus not treating it all to potable quality) offers significant opportunities in this respect, as well as delivering resource and cost savings. The fact that Europe, despite several initiatives, does not yet have a unified regulatory regime which can boost the embryonic reuse sector and protect citizens' interests is, from our perspective, disappointing. Compared with other countries (e.g. Australia and the US), European progress has been unhurried and lacklustre. Guidance and standards for non-potable reuse schemes are perhaps unsurprisingly more commonly available than those for potable applications, with several countries (e.g. Greece and Spain) having mature and comprehensive regulations. The argument developed in this article is in many respects intended as a challenge to politicians and regulators. A decision on how integration and subsidiarity should be balanced with respect to water reuse regulation for EU member states is overdue. Whether the context is potable or non-potable applications, the challenge is the same – a socially and economically profitable European water reuse sector requires the direction and confidence of a progressive and enabling regulatory regime.

One might argue that there is little urgency to develop such governance tools whilst other interventions remain viable and capable of making significant contributions to the supply–demand balance. This is a valid argument and has perhaps influenced the observed (lack of) pace and resolve to date. However, we would argue that under conditions where none of the component trends of the supply–demand balance are moving in a useful direction, the time has come to inject some urgency into the process. Regulation which is catalyzed by a crisis is rarely good regulation, but regulation informed by an appreciation of changing circumstances can drive innovation and provide the confidence which emerging actors need to plan and resource their initiatives. The nascent European water reuse sector, recently emboldened by the founding of its own industry association (Water Reuse Europe), can only grow and make a meaningful contribution to a sustainable water future for the region if there is progressive enabling legislation in place to frame its initiatives and operations.

Disclosure statement

No potential conflict of interest was reported by the authors.

References

Abwasserverband, Braunschweig. (n.d.). Company's website. Retrieved from http://www.abwasserverbandbs.de/en/what-we-do/the-braunschweig-model/

Alcalde Sanz, L. & Gawlik, B. (2014). *Water reuse in Europe - Relevant guidelines, needs for and barriers to innovation*. Report of the Joint Research Centre (NoJRC92582) JRC. Retrieved from http://publications.jrc.ec.europa.eu/repository/handle/JRC92582 .

Almeida, M. C., Vieira, P., & Smeets, P. (2014). Extending the water safety plan concept to the urban water cycle. *Water Policy.*, *16*, 298–322.

Asano, T. (1998). *Wastewater reclamation and reuse*: Water Quality Management Library Series, vol. 10. Lancaster, Pennsylvania: Technomic Publishing Co. Inc. 1528.

Bakker, K. (2003). *Good Governance in restructuring water supply: A handbook* (p. 44). Ottawa: Federation of Canadian Municipalities.

Bartram, J., Corrales, L., Davison, A., Deere, D., Gordon, B., Howard, G., Rinehold, A., & Stevens, M. (2009). *Water safety plan manual: Step-by-step risk management for drinking - water suppliers* (p. 101). Geneva: World Health Organization.

BIO by Deloitte. (2015a). *Optimising water reuse in the EU* – Final report prepared for the European Commission (DG ENV), Part I. In collaboration with ICF and Cranfield University. Luxembourg: Publications Office of the European Union, 45.

BIO by Deloitte. (2015b). *Optimising water reuse in the EU* – Public consultation analysis report prepared for the European Commission (DG ENV). Luxembourg: Publications Office of the European Union, 45.

Casiraghi, M. (2014). *Processi di depurazione urbane. "Processi Depurazione Acque Urbane Con Visita Ad Impianto Di San Rocco-Comune Di Milano"* workshop, October 2014, Milan, Italy. Retrieved from https://www.ordineingegneri.milano.it/fondazione/eventi-passati/depurazione-acque-urbane-san-rocco/casiraghi_OrdineIngg.pdf

EC. (1998).Council Directive 98/83/EC of 3 November 1998 on the quality of water intended for human consumption. *Official Journal of the European Communities*, L 330/32. Retrieved from http://eur-lex.europa.eu/legal-content/EN/TXT/?uri=CELEX:31998L0083

EC. (2000). Directive 2000/60/EC of the European Parliament and of the Council of 23 October 2000 establishing a framework for Community action in the field of water policy. *Official Journal of the European Communities*, L327. Retrieved from http://eur-lex.europa.eu/legalcontent/EN/TXT/?uri=CELEX:32000L0060

EC. (2006). Directive 2006/118/EC of the European Parliament and of the Council of 12 December 2006 on the protection of groundwater against pollution and deterioration. *Official Journal of the European Union*, L 372/19. Retrieved from: http://eur-lex.europa.eu/legal-content/EN/TXT/PDF/?uri=CELEX:32006L0118&from=EN

EC. (2013). Directive 2013/39/EU of the European Parliament and of the Council of 12 August 2013 amending Directives 2000/60/EC and 2008/105/EC as regards priority substances in the field of water policy. *Official Journal of the European Union*, Retrieved from http://eur-lex.europa.eu/legal-content/EN/TXT/?qid=1448032144816&uri=CELEX:32013L0039.

EEC. (1991).Council Directive 91/271/EEC concerning urban wastewater treatment. *Official Journal of the European Communities*, L135/40. Retrieved from http://eur-lex.europa.eu/legal-content/EN/TXT/?uri=CELEX:31991L0271

European Commission. (1999). *Approximation of Environmental legislation. Role of Compliance Costing for Approximation of EU Environmental Legislation in Cyprus*. A Study for the European Commission DG – XI.A.4, Annexes, June 1999. 36pp. Retrieved from http://ec.europa.eu/environment/archives/international_issues/pdf/cocyp_annex.pdf.

European Commission. (2013). *A water blueprint for Europe*. Luxembourg: Publications Office of the European Union.

European Commission. (2014). *Background document to the public consultation on policy options to optimise water reuse in the EU*. Responsible services : Unit ENV/C/1 Water, DG ENV, European Commission. Retrieved from http://ec.europa.eu/environment/consultations/water_reuse_en.htm

European Environment Agency. (2009). *Water resources across Europe – confronting water scarcity and drought*. Report No 2/2009, European Environment Agency, Copenhagen.

Godfree, A., & Godfrey, S. (2008). Water reuse criteria: environmental and health risk based standards and guidelines. In B. Jiménez and T. Asano (Eds.), *Water reuse: An International Survey of current practice, issues and needs* (pp 351–369). London: IWA Publishing.

Goodwin, D., Raffin, M., Jeffrey, P., & Smith, H. (2015). Applying the water safety plan to water reuse: Towards a conceptual risk management framework. *Environmental Science: Water Research & Technology., 1*, 709–722.

Hochstrat, R., Wintgens, T., & Melin, T. (2008). Development of integrated water reuse strategies. *Desalination, 218*, 208–217.

Hochstrat, R., Wintgens, T., Melin, T., & Jeffrey, P. (2006). Assessing the European waste water reclamation and reuse potential – a scenario analysis. *Desalination, 188*, 1–8.

Knight, H., Maybank, R., Hannan, P., King, D., & Rigley, R. (2012). *Learning legacy: The old ford water recycling plant and non-potable water distribution network*. Retrieved from http://learninglegacy.independent.gov.uk/documents/pdfs/sustainability/old-ford-case-study.pdf

Larcou Yiannakou, A. L. (2012). *Treated effluent reuse scheme in Cyprus*. Workshop on the quality of recycled water and its application in agriculture organised by the sewerage board of Limassol – Amathus, Limassol, April 2012. Retrieved from http://www.sbla.com.cy/sala_website/PRESENTATIONS/Mrs%20Angeliki%20Larkou%20-%20sbla%20treated%20effluent%20reuse%20scheme%20in%20Cyprus.pdf

Loubier, S., & Declercq, R. (2014). *Analyses coûts-bénéfices sur la mise en œuvre de projets de réutilisation des eaux usées traitées (REUSE) Application à trois cas d'études Français. Final Report (in French)*- - Onema – Irstea, 37pp. Retrieved from http://www.onema.fr/IMG/pdf/2014_025.pdf

Mazzini, R., Pedrazzi, L., & Lazarova, V. (2013). Production of high quality recycled water for agricultural irrigation in Milan. In V. Lazarova, T. Asano, A. Bahri, & J. Anderson (Eds.), *Milestones in water reuse: the best success stories* (pp. 179–190). London: IWA publishing.

Mujeriego, J., Sala, L., & Couso, R. (2011). Water reuse project of Tossa de Mar Consorci of Costa Brava. 8th IWA International conference on water reuse and reclamation, September 2011, Barcelona, Spain). Retrieved from http://www.ccbgi.org/docs/iwa_bcn_2011/Technicalvisit-WaterReuseProjectofTossadeMar.pdf

Nandha, M., Jeffrey, P., Berry, M., & Jefferson, B. (2013). Risk assessment frameworks for MAR schemes in the UK. *Environmental Earth Sciences., 73*, 7747–7757.

Nocella, G., Boecker, A., Hubbard, L., & Scarpa, R. (2012). Eliciting consumer preferences for certified animal-friendly foods: can elements of the theory of planned behavior improve choice experiment analysis? *Psychology and Marketing., 29*, 850–868.

NRMMC. (2008). *Australian guidelines for water recycling: Managing health and environmental risks (Phase 2): augmentation of drinking water supplies.* Canberra: Natural Resource Management Ministerial Council (NRMMC), the Environment and Heritage Council (EPHC) and the National Health and Medical Research Council (NHMRC). Retrieved from: http://www.environment.gov.au/system/files/resources/9e4c2a10-fcee-48ab-a655-c4c045a615d0/files/water-recycling-guidelines-augmentation-drinking-22.pdf

Otoo, M., Mateo-Sagasta, J., & Madurangi, G. (2015). Economics of water reuse for industrial, environemntal, recreational and potable purpose. In P. Drechsel, M. Qadlr & D. Wichelns (Eds.), *Wastewater: an economic asset in an urbanizing world* (pp. 169–192). New York, NY: Springer.

Papaiacovou, I., & Papatheodoulou, A. (2013). Integration of water reuse for the sustainable management of water resources in Cyprus. In V. Lazarova, T. Asano, A. Bahri & J. Anderson (Eds.), *Milestones in water reuse: the best success stories* (pp. 75–82). London: IWA publishing.

Paranychianakis, N. V., Salgot, M., Snyder, S. A., & Angelakis, A. N. (2015). Water reuse in EU States: necessity for uniform criteria to mitigate human and environmental risks. *Critical Reviews in Environmental Science and Technology, 45*, 1409–1468.

Roux, A., Robillot, C., Faye Chapman, H., Leusch, F., Hodge, M., & Walker, T. (2010). Hazard identification, qualitative risk assessment and monitoring on the Western Corridor Recycled Water Project. *Water Practice & Technology, 5*, 1–10.

Sala, L. (2010). *Evolution of water reuse practices in Tossa de Mar: from environmental reuse to urban non-potable reuse.* Workshop on water reuse, University of Taiwan, Taipei, November 2010. Retrieved from http://www.ccbgi.org/docs/taiwan_2010/l_sala_taiwan_2010.pdf

Salgot, M., Vergés, C., & Angelakis, A. N. (2003). Risk assessment in wastewater recycling and reuse. *Water Science and Technology: Water Supply, 3*, 301–309.

Sanchez-Flores, R., Conner, A., & Kaiser, R. A. (2016). The regulatory framework of reclaimed wastewater for potable reuse in the United States of America. *International Journal of Water Resources Development.* http://dx.doi.org/10.1080/07900627.2015.1129318.

Technopolis Group. (2013). *Screening of regulatory framework, Final Report November 2013.* Developed for the European Commission (DG-RTD C1), Project Acronym: 1770 'Regulatory Screening', Contract: MS (2012) 1055187. Retrieved from http://www.technopolis-group.com/wp-content/uploads/2014/07/1770-Final-report_13112013_edited-copy.pdf.

Ternes, T. A., Bonerz, M., Herrmann, N., Teiser, B., & Andersen, H. R. (2007). Irrigation of treated wastewater in Braunschweig, Germany: An option to remove pharmaceuticals and musk fragrances. *Chemosphere, 66*, 894–904.

Van Houtte, E., Cauwenberghs, J., Weemas, M., & Thoeye, C. (2012). Indirect potable reuse via managed aquifer recharge in the Torreele, St-André project. In C. Kazner, T. Wintgens, & D. Dillon (Eds.), *Water reclamation technologies for safe managed aquifer recharge* (pp. 33–46). London: IWA Publishing.

Van Houtte, E., & Verbauwhede, J. (2012). Sustainable groundwater management using reclaimed water: the Torreele/St-André case in Flanders, Belgium. *Journal of Water Supply: Research and Technology - AQUA, 61*, 473–483.

WHO. (2006). *WHO Guidelines for the safe use of wastewater, excreta, and greywater. Volume I-IV.* Geneva: World Health Organisation. Retrieved from: http://www.who.int/water_sanitation_health/wastewater/gsuww/en/

WWF. (2010). *Riverside tales: Lessons for water management reform from three English rivers.* WWF UK. Retrieved from: http://assets.wwf.org.uk/downloads/riverside_tales.pdf

Policy issues confronting Australian urban water reuse

James Horne

College of Asia and the Pacific, Australian National University, Canberra, Australia

ABSTRACT
Urban water security in Australia's major cities is now very high, reflecting in part recent policy interventions. Important indirect potable water reuse projects were completed but no direct potable reuse project was undertaken and none seems likely in the near term. Governments have much to learn from decisions to build very large desalination and recycling plants, particularly around timing and scale. Future water reuse decisions are likely to have a much greater commercial focus. Policies and regulations giving more flexibility to decentralized provision of water-related services could result in further growth of climate-resilient water resources and non-potable reuse.

Introduction

The Millennium drought (from the late 1990s to around 2009) saw quite fundamental change in water management in Australia. Deep introspection took place not only in water management in rural and regional areas but also in Australia's urban areas (Horne, 2013). Established backward-looking approaches to surface water inflow modelling were supplanted by the need to consider future climate impacts on inflows, rather than examining historical experience (Melbourne Water Climate Change Study, 2005; NSW Office of Water, 2010; Paton, Dandy, & Maier, 2014). Reduced water availability in all the mainland capital cities and many regional urban centres put a much greater focus on urban water reuse, and climate-resilient forms of supply augmentation. With around 90% of its population living in urban areas and 61% in six of its largest cities, urban water availability figured prominently in national, state and community discussions on how the water availability crisis should be managed.

This article reviews policy issues confronting urban water reuse, including direct potable reuse (DPR) and industrial water reuse, in the context of demand and supply options to manage urban water security. Its purpose is to understand the potential role of DPR and industrial water reuse in future. It takes as given that the policy framework and resource management should consider climate change impacts.

Unlike the United States, where DPR projects are gaining momentum in states confronting serious water scarcity (Raucher & Tchobanoglous, 2014), there are no DPR projects operating in Australia. However, as can be seen from the Australian Bureau of Meteorology's (BOM)

fledgling website entitled 'Climate Resilient Water Sources' (Bureau of Meteorology, 2015a), water reuse and desalination projects, including urban water reuse projects, are very much an established feature of the Australian landscape. The BOM site currently covers 360 sites Australia-wide (with over 260 relating to water reuse). In 2012–13, production capacity was put at 1820 GL/year and production at 440 GL, covering facilities from small remote urban centres and remote industrial users to the major urban centres and food and beverage industrial facilities. Table 1 lists some significant examples of water reuse projects and their key attributes.

As is the case in many countries, de facto indirect potable reuse has been a feature of urban supply for many inland urban areas for some time (GHD, 2007). Indeed, the national capital, Canberra, discharges the bulk of its treated high-quality wastewater into the Murrumbidgee River system, where it is used downstream by many small urban centres (Gregory & Hall, 2011). And an important indirect potable reuse scheme (IPR) involving managed aquifer recharge has commenced operating in Perth, Western Australia (Moscovis, 2013; Water Corporation, 2014b).

On the other hand, as late as 2006 the Water Services Association of Australia (WSAA) was on record as not supporting the introduction of DPR because of its risk relative to IPR alternatives (WSAA, 2006). Later still, two forward-looking reports on urban water (National Water Commission, 2014; Productivity Commission, 2011) make no mention at all of DPR. This is surprising as one of those reports is highly critical of governments ruling out, for political reasons, options relating to rural–urban transfers of water (Productivity Commission, 2011).

This article reviews urban water reuse history in Australia in the context of six major urban centres over the decade since the 2004 signing of the National Water Initiative (NWI) (Council of Australian Governments, 2004), in particular focusing on where the policy debate has come from and where it has reached. With the exception of Canberra, the other five are coastal state capital cities, namely Sydney, Melbourne, Brisbane, Perth and Adelaide, with populations of between 1 million and 5 million inhabitants. In total the population of these cities is currently around 14 million. In each case, the state government and local councils play the key regulatory and policy roles in relation to drinking water distribution and wastewater, with the role of the national government focusing on driving reform through national guide-lines around water pricing and water quality/health-related issues, capacity-building and information, and consumer advice (Australian Government n.d.; Productivity Commission, 2011). This article then examines potential future directions for urban water reuse policy in Australia with specific reference to DPR and industrial reuse in the context of managing urban water security. Finally, conclusions and lessons are drawn for both Australian and international audiences.

Urban water reuse history

The 2000s saw a significant change in attitudes and approaches to urban water. While some states already had in place programmes to increase urban water reuse, in 2003 the national and state governments commenced an initiative to update the existing water management framework resulting in an intergovernmental agreement covering both rural and urban water issues (Council of Australian Governments, 2004). On urban water, the NWI noted that the national and state governments sought:

Table 1. Significant Australian water reuse exemplars.[a]

Project	Type	Date	Capacity	Comment
Perth Groundwater Replenishment (Water Corporation)	Potable reuse scheme (IPR)	2014	14 GL	Exemplifies large-scale centralized IPR that effectively meets drinking water standards. See Table 3 for details
Flow Systems local water utility	Non-potable reuse (utilizing stormwater, rainwater and wastewater)	2014	Contracted to increase to 10 ML/day	Exemplifies the positive impact of competition in the provision of decentralized water and wastewater services. Manages water and wastewater services for eight communities (22,000 dwellings), including Central Park Water and Green Square Water in Sydney
Western Corridor Recycling (SE Queensland)	IPR/non-potable recycling	2008–10	80 GL/year	Large centralized potable reuse project; pricing not properly resolved; mothballed in 2013; previously provided three power stations with recycled water
K2 sustainable housing project	Non-potable reuse	2008	3 ML/year	Exemplifies modern water management in housing development. Reduce potable demand by > 50%.
Coca Cola plant Richland	Minimizing discharge	2004	4 ML/day	95% of incoming water is used in production; remaining wastewater is reused in irrigation
Arnotts Snackfood	Non-potable reuse	2011	106 ML/year	Uses membrane filter technology to treat and recycle potato-washing water before using it in industrial processes
1 Bligh St Sydney (Aquacell)	Non-potable reuse		22 ML/year	Exemplifies modern decentralized water management in a new office block
Yatala Brewery (Osmoflo)	Reuse water replacing potable supply	2004	4 ML/day	Exemplifies the role of integrated management of drinking water and reuse water; allowed a doubling of production capacity using only 15% more water
Cooper's Brewery Adelaide (Osmoflo)	Replacing potable supply	2007	3 ML/day	Exemplifies reverse osmosis applied to groundwater
Sydney Water recycling for industrial purposes (Sydney Water)	Eighteen plants supporting 23 recycling schemes	Various	23 (of 47) GL in 2013–14	Recycled water comprises 9% of bulk water supplied. Industry uses around half; steel production, coal loaders and wastewater plants are the prominent users

[a]These examples have been selected as showcasing aspects of water reuse. BOM (2015) lists 264 sites. WSAA (2009) provides 34 examples (two of which are included above). The Metropolitan Water Directorate (2015) lists 48 industrial-use projects relying on grey water, roof water, sewage – indirect, sewage – treated, sewage – untreated and stormwater in the Sydney Metropolitan area.
Sources: Aquacell Water Recycling (2015); Flow Systems (2015); Marsden Jacob Associates (2014); Osmoflo (n.d.); Sydney Water (2014); Water Corporation (2014a); WSAA (2009).

- Healthy, safe and reliable water supplies.
- An increase in water use efficiency.
- Reuse and recycling of wastewater where cost-effective.
- To facilitate water trading between urban and rural sectors.
- To encourage innovation throughout the urban water cycle.
- Adherence to full cost recovery, and to develop pricing policies for recycling and stormwater to stimulate efficient use (Council of Australian Governments, 2004).

The outcomes sought indicated concerns across the Australian urban water landscape, requiring many entrenched water policy problems to be considered. Australia's large urban water utilities, which had operated largely as comfortable city-based monopolies, were being put under the microscope. Actions agreed by the state and national government included demand management strategies, innovation and capacity-building to create water-sensitive

Table 2. Options to strengthen urban water security.

Option	Water security	Issues	Examples
Increase supply			
Dam augmentation	Case by case as water use in catchments has largely reached or surpassed sustainable extraction levels. In some catchments climate change has already reduced inflows significantly	Major environmental issues. Established technology with few risks that are not well understood and accepted	Canberra's Cotter Dam rebuilt to increase capacity from 4 to 76 GL, adding 35% to Canberra's overall storage network. Several dam proposals were rejected on environmental grounds
Rural–urban transfers	Viable option to bolster water security (Quiggin, 2006) in Melbourne, Adelaide and Canberra	Rural–urban transfers hampered by political constraints in Melbourne and Adelaide notwithstanding 2008 National Urban Water Planning Principles exhorting the consideration of all options (Australian Government n.d.)	The 70 km 'North South' pipeline was constructed from the Goulburn River to Melbourne's water storages. It has an annual carrying capacity of 75 GL, but is currently not used (Melbourne Melbourne Water, 2011), reflecting political intervention
Increase groundwater extraction	If use exceeds sustainable yield this will detract from water security; new technologies to delineate groundwater resources are creating opportunities to service small urban areas	Improving desalination and reverse osmosis technologies, and lower costs	Opportunities to upgrade water security for small urban centres abound
Rainwater capture through rainwater tanks	Rainfall dependent; at best a supplementary source	Suitable for potable and non-potable uses. Cost-effectiveness and quality control are key issues	Millennium drought saw household tank capacity in capital cities grow by 80%, adding 20 GL to supply
Desalination of seawater/brackish groundwater	Climate resilient. Increases water security. Resource infinite for appropriately placed coastal cities. Also useful for small inland centres	Cost-effectiveness the key issue, but highly cost-effective in some markets, such as Perth. Construction decisions reflected politics rather than compelling business cases	Six large plants with a capacity around 480 GL were constructed. A majority of these plants are currently mothballed. Perth's two plants are the exception. In 2013–14 they provided 113 GL, or 39%, of its Integrated water scheme supply
Water reuse for non-potable purposes, particularly for industrial use	Increases water security; currently a significantly underutilized resource that is for the most part discharged into the ocean	Cost-effectiveness a key issue, but significant take up in recent years. The Western Corridor Recycled Water Project (SE Queensland) is mothballed due to lack of demand	Each city has numerous examples involving industrial reuse of wastewater from wastewater treatment plants. For example, Rosehill (Sydney)
Water reuse for non-potable purposes, particularly commercial and residential use in new large-scale infill developments	Increases water security; currently a significantly underutilized resource that is for most part discharged into the ocean	Ensuring competition policy rules and resource ownership are well defined. Costs often hidden in planning regulations need exposing	Many inner-city developments involve devolved plant, recycling the development's wastewater or mining proximate sewers for resources. Examples include 1 Bligh St Sydney and NAB Bourke St campus Melbourne. See also the flow systems model
Indirect potable reuse	Increases water security	Water quality to meet 2004 Australian Drinking Water Guidelines (ADWG). Technology is established, involving environmental buffers as a last step. Key issue around community acceptance in some states and communities	Perth is currently constructing a potable reuse scheme (IPR) scheme based around groundwater replenishment, which when complete will provide around 10% of Perth's water supply. See Table 3 for details

(Continued)

Table 2. (*continued*).

Option	Water security	Issues	Examples
Direct potable reuse	Increases water security	Water quality to meet ADWG. Technology not yet fully accepted in all states or by the community. Key issue around community acceptance	None in operation, or currently being contemplated
Reduce demand			
Appropriate (high-er) prices	Full cost recovery and a return on investment encourages reduced use; encourages appropri-ate maintenance. Both increase security	Basic policy well accepted in major urban areas, but national pricing principles not yet fully implemented	All cities have prices set by inde-pendent economic regulators; significantly higher prices have reduced demand
Competition poli-cy framework	Enhances short-term water security as competition intensifies		New legislation allowing decentralized solutions in new developments. See New South Wales's Water Industry Compe-tition Act (WICA) legislation
Water-efficien-cy labelling schemes promoting technological change	Reduces demand, promoting water security. Annual savings estimated at 70 GL/year	Well-accepted driver of water-saving consumer attitudes (Fyfe et al., 2015; Guest, 2010)	Water Efficiency Labelling and Standards (WELS) Scheme, an ongoing cost-effective scheme to promote the adoption of water-efficient technologies, e.g. around showerheads, toilets and washing machines
Mandating water-use out-comes in new developments	Reduces demand but may not improve 'security' as it is already taken into account in planning	Hidden costs	State governments and local councils planning rules. See the City of Sydney water management plan
Regulatory restric-tions	May reduce security if processes are ad hoc or discretionary	Hidden costs (Grafton & Ward, 2008) without a long-term effect on water conservation values (Quan-tum Market Research, 2014)	Outdoor water restrictions a feature of the drought period, resulting in significant unac-knowledged costs

Sources: ABS (2014); Abrams et al. (2011); Aquacell Water Recycling (2015); Flow Systems (2015); Fyfe et al. (2015); Guest (2010); National Water Commission (2014); seqwater (2015); Water Corporation (2014b); Western Corridor Recycled Water Pty Ltd (2009).

cities. For the first time, the Council of Australian Governments (COAG) commissioned the development of national guidelines for health and environmental issues covering recycled water and stormwater (COAG, 2004).

In 2005, urban water reuse focused heavily on reuse in the agricultural sector and urban parklands, with minor volumes of recycled water being used by commercial, industrial and residential users (ABS, 2006). Reflecting the onset of the Millennium drought, and nascent policy reforms, 'water savings' in the largest urban areas (Sydney and Melbourne) had just started to be taken seriously (Gregory & Hall, 2011). Managing water security (here defined loosely as 'the availability of an acceptable quantity and quality of water for health, liveli-hoods, ecosystems and production'; Grey and Sadoff, 2007, p. 545) was the major policy driver.

While not the first example, the failed Toowoomba indirect potable reuse proposal in 2005–06 illustrated that while de facto potable reuse was and still remains an important element of water resources for many communities, attempts to introduce such schemes

Table 3. Potable reuse scheme (IPR) for Perth's groundwater replenishment.

- Project provides major insights into water reuse to produce products that meet 2004 Australian Drinking Water Guidelines (ADWG), which is then to be stored in groundwater aquifers for future use
- The trial, which preceded the major investment in supply augmentation, took nearly 10 years, working meticulously through policy, technical and community consultation issues
- At a policy level, the Western Australia Department of Health was responsible for public health, the Department of Water for protecting water resources, and the Department of Environment and Conservation for protecting the environment
- A key policy issue was that water produced from an advanced water recycling plant (AWRP) was defined as wastewater, regardless of the water quality manufactured. An approach was required to develop new health and environmental regulations to allow water produced by the AWRP to allow groundwater replenishment, and new policy rules for allocation of replenished groundwater
- The Department of Health study on water quality developed 254 water quality guidelines that must be met at the point of recharge
- The first stage of the groundwater augmentation scheme is scheduled to be operational in 2016, with a capacity to recharge 14 billion litres of water into groundwater aquifers (Moscovis, 2013; Water Corporation 2014b)

knowingly, or in a planned way, could run into issues around community acceptance, no matter what the scientific basis of the proposal (Hurlimann & Dolnicar, 2010).

The 2004 Australian Drinking Water Guidelines (ADWG) constituted Australia's risk management framework directed largely at traditional water resources, such as dams and bore water (NHMRC & NRMMC, 2014). Recognizing the thrust of the NWI and the events of Toowoomba, the Australian Guidelines for Water Recycling (AGWR) Phase 1 were introduced in 2006 and directed at ensuring non-potable recycled water was of an appropriate quality, 'promot(ing) a quantitative assessment of health-based risks, with a strong focus on risks from pathogens' (ATSE, 2013, p. 80). In a policy sense, these guidelines provided a national framework for a safe expansion of urban water reuse, while not entering into debates or issues around DPR.

AGWR Phase 2 was released in 2008, covering augmentation of drinking water supplies (EPHC, NHMRC, & NRMMC, 2008), and stormwater use and managed aquifer recharge (NRMMC, EPHC, & NHMRC, 2009). The module covering augmentation of drinking water supplies, which specifically relates to 'the use of recycled water to supplement drinking water supplies', discusses DPR quite explicitly:

> Direct augmentation should not proceed unless sufficient mechanisms are established to prevent substandard water from being supplied. Implementation of direct augmentation presents substantial technical and management challenges. The need for reliability of processes, vigilance of monitoring and highly skilled operators – already high for indirect use – is magnified for direct augmentation. Knowledge and understanding of system reliability and control of variability is essential before direct augmentation can proceed. Further research is required in this area. (EPHC, NHMRC, & NRMMC, 2008, p. 4)

As policy decisions on the health aspects of the regulatory guidelines were the province of state and local government, no state government would in the view of the author countenance DPR against this advice.

By 2007 the national government committed itself to a target of recycling 30% of wastewater by 2015, well above both target rates already committed to by state governments and the average recycling level of around 9% (Khan, 2008; Marsden Jacob Associates, 2008; WSAA & NWC, 2009), effectively signalling it was prepared to underwrite new capacity to lift the rate of recycling. There was no specific rationale for the 30% target; indeed, it seems to have been based on being seen to respond to the deepening water availability crisis, and developing a political platform for the earlier national election (Albanese, 2006). A report commissioned after the target had been established to test its achievability found that the overall rate of recycling could increase to 23% by 2015, but would not reach 30% without massive additional investment (Marsden Jacob Associates, 2008). Lack of a good-quality data series makes progress difficult to judge, but there appears to have been little growth in water reuse of in recent years (Bureau of Meteorology, 2015c).

Over the second half of the decade the thrust of urban water policy in each city was to strengthen water security. The supply and demand options are familiar, and set out in Table 2.

Consumption per capita fell through much of the period 2000–10, driven by policy initiatives that sought to reduce demand through increased water prices, raised consumer awareness of water efficient technology (e.g., water-efficient taps and shower heads) and mandatory restrictions on use during the drought years. These policy initiatives were reinforced by a growth in apartment living, and to some degree changing attitudes towards water use. Water prices increased significantly in real terms off the back of inclining block tariffs adopted by urban utilities (National Water Commission, 2014). Empirical evidence suggests that this reduced demand by a modest amount, but the price impact may have been reduced by upgrading of appliance efficiency (Abrams, Kumaradevan, Sarafidis, & Spaninks, 2011).

Currently, water security and water quality in major urban centres is very high (Bureau of Meteorology, 2015b; National Water Commission, 2014). State and local governments and their major water service providers in each of the major centres have grappled with water security problems over the past decade and made some large (and sometimes very expensive) investment decisions covering recycled water and desalination plant capacity (Byrnes, 2013). Notwithstanding a 2011 Productivity Commission report critical of many aspects of past urban water policy (Productivity Commission, 2011), Australia's major urban centres start with these decisions behind them, and the resultant infrastructure in place. The additional water production capacity from desalination plants alone would be more than enough to meet the additional water demand from the population increase of the six urban areas growing by 2% per year to 2030, assuming a generous consumption per capita of 250 litres/day.

For this reason, as of mid-2015, the pressure on governments and urban water policy is perhaps lower than at any time in the past decade, notwithstanding the recommendations of the Harper Competition Policy Review (which, *inter alia*, presented the national government with a package of proposals to reform Australia's competition policy and law, including aspects applying to water; Harper, Anderson, McClusky, & O'Bryan, 2015). The large investments in desalination and recycling infrastructure of half a decade ago are sunk costs. Notwithstanding efforts to raise its profile (ATSE, 2013; Institute for Sustainable Futures, 2013), DPR and urban water reuse policy are currently not high on the political agenda at either state or national level. Legislative and regulatory arrangements (NSW Government, 2014) are shaping outcomes, and major urban water service providers continue to strive to meet their corporate goals (Melbourne Water, 2014; Sydney Water, 2014; Water Corporation, 2014a). National leadership through COAG has all but disappeared with the abolition of the ministerial councils covering water issues (Council of Australian Governments, 2013). However, urban water supply services are being delivered by a gradually increasing number of bodies working to use available resources more effectively (Chong, 2014).

Future directions in urban water reuse policy in Australia

A number of policy issues are likely to be important in any discussion of future directions in urban water reuse policy in Australia. These include the following, which are discussed below: competition, access and pricing; water quality and health; community engagement; climate change; and interaction with other policies such as integrated urban planning.

Notwithstanding the desalination/water reuse capacity overhang, a key issue for policy is ensuring that investors (be they from the public or private sector) can make cost-effective decisions that well reflect core objectives around water security, while providing competitive prices to users.

Effective consideration of these issues by governments will require good underlying science and information that can be readily accessed. Cooperative research centres (including the Australian Water Recycling Centre of Excellence – AWRCOE, the National Centre of Excellence in Desalination and the Cooperative Research Centre for Water Sensitive Cities) service aspects of this need for defined periods, while national bodies such as the BOM, dominant water supply providers and state agencies can develop and maintain databases that should reduce transaction costs over time and improve the depth and breadth of information availability.

Competition and pricing policy

The structure of the market and issues around entry, exit and pricing can have a large impact on how new approaches to water manufacturing or accessing new water resources are managed. These issues have been managed quite differently city to city (Byrnes, 2013; Chong, 2014; Humphries, 2014; Productivity Commission, 2011; Samuel, 2014). Some remain at the centre of contemporary debate (Harper et al., 2015).

On competition policy, New South Wales has led the way with its Water Industry Competition Act (WICA) arrangements. Year 2014 amendments to the original 2006 legislation (expected to commence in 2016) have provided an enhanced basis for competition from private sector service providers to the incumbent dominant state owned service provider (Flow Systems, 2014; Humphries, 2014). WICA has facilitated privately financed water reuse projects. The 2014 amendments can be expected to reduce barriers to entry while at the same time maintaining strong regulatory oversight of water quality issues. However, paradoxically, these new arrangements are unlikely to result in an early positive impact on DPR diffusion. Indeed, it may well delay it because they allow industry service providers to conceive and implement new business models that utilize existing potable water supplies for drinking water, buttressed with new approaches to all other water demand needs and to wastewater/water reuse services (Aquacell Water Recycling, 2015; Flow Systems, 2015). This encourages price competition in service provision and greater water reuse where it is cost-effective. Moreover, while some (Marsden Jacob Associates, 2013) see legislative and regulatory barriers as being largely resolved, new entrants like Flow Systems see further room for regulatory reform to encourage greater efficiency and clearer division of roles between government agencies and the dominant water service provider (Flow Systems, 2014). With significant urban infill projects being undertaken in Sydney, which also has considerable ageing wastewater infrastructure, the list of new projects involving water reuse will continue to grow.

New South Wales's competition policy approach to encourage private investment is quite different to that adopted by state governments responsible for other major cities, which generally are still protective of the traditional government corporation business model to deliver water and wastewater services, even although all states now have an independent economic regulator or water pricing authority. Pricing regimes implemented by independent regulators in each state and the Australian Capital Territory (ACT) seek to recover at least full costs, and all the state-owned enterprises are profitable (Samuel, 2014). Support for the NWI pricing principles increases the prospect of a thorough consideration of water reuse options to satisfy some part of an urban area's emerging water demand, and to non-potable recycling in some forms. However, with the exception of Perth, such an approach seems unlikely to result in DPR capacity in the foreseeable future.

The 2015 Harper Competition Policy Review has argued governments should strengthen the national regulatory focus on water, 'creating incentives for increased private sector participation in the sector through improved pricing practices' (Harper et al., 2015, p. 205). The review argues the key 2004 NWI urban water objectives have not been met on a nationally consistent basis, and that consistency would encourage long-term private investment (Harper et al., 2015). There is little doubt such an approach would encourage decentralized solutions to urban water reuse by reducing transaction costs.

There has been some debate about the potential role of 'flexible' or 'scarcity' pricing, largely in the context of the bulk water provision, to reflect better temporal changes in the supply demand balance (Frontier Economics, 2008; Productivity Commission, 2011). But it is not under consideration by any government. There is an ongoing discussion around use and pricing of stormwater and wastewater services (IPART, 2014; Melbourne Water, 2013). These are complex issues for governments and the economic regulators to grapple with, but can have an important bearing on the extent to which and how quickly the water reuse industry can grow if circumstances warrant that.

In the case of stormwater and wastewater services, the service provider is often local government, requiring careful attention to detail in the regulation framework. The debate in this area is complex and the analysis affected by issues such as clarity around the property right framework, institutional arrangements, and legislation on roles and responsibilities. Technology is already providing opportunities for decentralized urban water reuse. A key issue for policy is to consider arrangements that both lower costs and prices and enhance water security while at the same time ensuring an appropriate return on infrastructure investments.

Desalination and reuse technologies are continuing to challenge the notion of vulnerability to the variability of climate as it applies to Australia's major cities. Instead, manufactured water – be it potable or non-potable water for drinking, industrial and other uses – has the ability to transform these notions. However, barriers to new entrants remain (Flow Systems, 2014; Marsden Jacob Associates, 2013). The policy frameworks around competition, property rights, role of government corporations, pricing and water quality will play a role in how these opportunities are utilized. Outside of New South Wales much greater focus is required on access and competition.

Community engagement

Issues relating to community acceptability have surrounded projects involving IPR and DPR. Handled poorly, e.g. in the case of the Toowoomba IPR project (Hurlimann & Dolnicar, 2010), community reaction can result in cancellation of a proposed reuse project. Handled well, e.g. as in the case of Perth's groundwater replenishment, it can buttress support for new projects and provide a solid basis for expanding potable reuse (Marsden Jacob Associates, 2014; Moscovis, 2013). Actions to improve public acceptance of DPR include developing messages and information to address safety and water quality concerns, commencing communications before a specific DPR project is being considered, and facilitating improved community understanding via education initiatives (ATSE, 2013). The Australian Water Recycling Centre of Excellence (AWRCOE) has been engaged in developing a National Demonstration, Education & Engagement Program in relation to Water Recycling for Drinking (AWRCOE, 2015). As valuable as these programmes are, in that they will help address basic community confidence and acceptance in relation to certain water products, they are necessary but not sufficient conditions for DPR projects to proceed. The sufficient conditions go to the issue of its contribution to water security and cost-effectiveness. As noted above, whether both necessary and sufficient conditions will be met in the Australian urban environment in the medium-term will depend on the business models that emerge in coming years and their cost-effectiveness, and the ability and willingness of government to regulate to allow and encourage these business models.

Water quality and health policy

Meanwhile, developmental work continues in national health forums to ensure the risk framework around water reuse for potable purposes reflects and encompasses the latest scientific advice and technological advances respectively. At the level of scientific advice on safety of recycling, the main discussion is around microbial safety (ATSE, 2013). But while technology has progressed to allow effective management of risks to health from DPR, the issue of microbial safety has not yet been incorporated into the ADWG, even though coverage of it is incorporated in the 2008 AGWR. The Water Quality Advisory Committee of the National Health and Medical Research Council (NHMRC) (the national advisory body that updates the ADWG) is scheduled to undertake the requisite work by 2016.

Currently there do not appear to be any barriers in the AGWR or the ADWG to the introduction of DPR underscored by the fact that the Perth advanced water treatment plant for groundwater replenishment scheme trial was completely successful in meeting the stringent health guidelines for drinking water before the water was placed into the storage aquifer. The Western Australian Department of Health established some 254 water quality guidelines that needed to be met at the point of recharge (Moscovis, 2013). However, that project was always conceived as an IPR scheme. The language of the current AGWR is such that state government health advisors and politicians are unlikely to proceed under the current water quality guidelines.

Ongoing work surrounds the issue of a national validation (NatVal) framework for water reuse treatment technologies. Detailed nationally funded research has been underway for several years under the auspices of the AWRCOE (AWRCOE, 2015b; Aither, 2013). The work covers the perspective of regulators, technology suppliers and companies involved in water reuse (AWRCOE, 2015b). Trial pilots using the draft protocols have yet to commence. In recent months several state agencies have expressed some measure of support for the project, but there is nothing in the public domain. At the time of writing, while progress has been substantial, the underlying research and protocol development is incomplete. Without doubt this is important work but with the AWRCOE future uncertain beyond late 2016 what will happen to work undertaken thus far under the NatVal project auspices is unclear.

Climate change

Operating in a water resource framework recognizing climate change, water availability can no longer be operated on the assumption of stationarity (Milly et al., 2008). As noted above, major urban water supply providers have adopted new modelling based on climate change science and forward-looking estimates of the productivity of urban catchments in an attempt to ascertain future demand and supply gaps. But with the development of climate-resilient water sources and the prospect of small, decentralized water service providers, developing integrated frameworks encompassing climate change and those new water sources will be required. Policy needs to be reframed so as not to impede these developments.

In some urban catchments climate change may result in greater rainfall variability, with implications for storage management and water quality (Short, Peirson, Peters, & Cox, 2012) that need to be managed. Both science and data are important (Horne, 2015). Reflecting the above, the national government in particular has invested in developing science-based centres of excellence focused on climate resilient technologies (e.g., the AWRCOE and the National Centre of Excellence in Desalination). The BOM has also increased its focus on data collection from climate-resilient water sources (Bureau of Meteorology, 2015a).

In a drying climate, governments have shown more concern to establish enhanced climate resiliency around water availability rather than over energy use. The Western Australia government, facing major rainfall based shortfalls in supply, has had the largest risk to manage, and has done most in a policy sense to develop the regulatory framework around expand climate resilient potable water sources, namely IPR.

A number of commentators have noted the connection between desalination and water reuse and their energy intensity as a proxy for their adverse impact on carbon emissions (Herndon, 2013). But this argument has not been absorbed into mainstream Australian policy. The more important perspective is to allow each project to speak for itself in terms of benefits and costs of specific approaches. Energy used in urban water service provision is less than 1% of primary energy demand, and while desalination is energy intensive it does provide an immediate fillip to water security in periods of drought. In a consistency sense, the issue from a climate change policy perspective is not the intensity of energy use of any technology per se, but total carbon content of energy consumed. Greenhouse emissions vary considerably from city to city (Cook, Hall, & Gregory, 2012). Moreover, all existing capital city desalination plants claim a 100% reliance on renewable energy or offsets through the purchase of renewable energy certificates (WSAA, 2013). Water reuse also opens the prospect for nutrient extraction (particularly phosphorous and nitrogen) if it becomes economic (Law & Hall, 2014).

Policy interactions

The Australian urban environment is a complex mix of regulatory interventions, evidenced by the significant differences in regulatory frameworks between states. There are many and varied advocates for a myriad of policy prescriptions, but there can be no presumption of optimality of prescription, and no evidence that the strands of government policy are consistently brought together. Different governments will consider the same issues, and often accord them different priorities. Some prioritize energy use; some have planning precepts that put industry in certain places in the urban form; and some reflect ownership/service provision arrangements (e.g., retention of public ownership over private ownership) that to others are not meritorious. Australia's urban areas illustrate a mix of these different priorities and approaches to risk. How water reuse emerges out of the mix will be determined by how governments embrace this mix of challenges.

Reuse by industry

Water policy generally and water reuse policy specifically can affect industrial reuse via all the channels identified above.

Historically, industrial users of recycled water were located around wastewater treatment plants. But in line with new business models noted above, this is changing as smaller reuse plants are now being located within industrial facilities and firms are producing their own 'made to order' reuse water.

At the level of industrial reuse within existing industrial facilities, companies are availing themselves of onsite technologies to reuse water many times. The reasons for this are manifold, but are essentially financial, from reducing costs to enabling expansion of factory production without overburdening existing wastewater infrastructure (for which they might be required to pay). Projects need to meet at least the usual internal rate of return hurdles to receive financing within companies' internal investment frameworks.

Other factors may be important, including perceived impacts on brand value where health and food-related products are concerned. In some industries, companies must still be alert to problems with water reuse for products that are to be exported. Recent comments suggest some countries still have prohibitions around such use (AWRCOE, 2014). It is not clear that this is a significant impediment to use.

Government can promote reuse through removal of unnecessary red tape that restricts reuse. The NHMRC work noted above would prepare the groundwork to ensure health and safety regulations do not unnecessarily impede potable reuse; it remains unclear whether there is an appetite for national regulations or a national validation framework for water reuse technologies. Until then it will be up to each individual jurisdiction to rework their regulatory framework reflecting updated water quality frameworks as they are introduced. Currently policy in some states does not appear to reflect the risk management framework embodied in the ADWG and the AGWR, and is quite prescriptive about reuse by industry.

Conclusions and lessons

Water reuse policy in Australia is in a state of reassessment. Governments took many large infrastructure investment decisions during a perceived crisis of water availability. Now, with large dams full in many coastal urban areas, much of this climate-resilient public sector infrastructure has been mothballed and is likely to remain so for a lengthy period. The costs of the earlier period of political intervention (rather than consistent policy intervention) will be borne by the communities for at least the next decade.

From today's perspective, where most urban areas have high levels of short- to medium-term water security, DPR is not likely to be given mainstream policy consideration from a 'centralized' water provider perspective in the near-term. Governments have much to learn from decisions to build very large desalination plants, particularly around timing and scale. Since then, water reuse decisions (in both private and public sectors) have been taken with a much greater commercial focus. In this environment, DPR options are unlikely to be immediately viable, particularly when other less confronting (from a community perspective) and more cost-effective options are available.

The introduction of improved competition and pricing policy frameworks, and updating of risk-based water quality guidelines to reflect improved technology, will strengthen incentives for investors to test decentralized business models utilizing reused water. In most centres, expansion in non-potable reuse is likely in the first instance as a part of the product mix of new small water service providers, but this will have a strong substitution effect leaving existing potable supplies to manage population growth and a reduction in use of climate dependent supply sources.

But the Perth experience of developing a solid policy framework around a groundwater replenishment scheme, with all the attributes of DPR but sold as an IPR scheme, illustrates that if and when required DPR is a technology that deserves to be considered along with the other demand and supply options. Considering DPR and IPR in this full list of possible options will allow investors to make fully informed decisions. While there is no reason for DPR to be left off such a list of possible options, the fact that it was in both the 2011 Productivity Commission report and the 2014 NWC report suggests that at that time the institutions considered DPR a 'bridge too far' (National Water Commission, 2014; Productivity Commission, 2011). To be consistent with their general thrust that governments should consider all options, both reports should have explicitly made this case, as to not make the case suggests to governments that it is acceptable that other options also be left off the list (e.g., rural–urban transfers).

The Australian experience suggests 10 possible lessons for international consideration.

- Impacts from climate change mean that approaches to urban water security need to change (as well as other forms of water security) and become forward looking and risk based. Water resource managers may need to reassess floods and scarcity-related risk profiles, and their impact on urban water security.
- Technology developments suggest new ways are emerging to address possible shortfalls in urban water availability through increased reliance on climate-resilient water resources. To take advantage of these developments the responsible urban governments will need to examine carefully the case for revamping competition policy and encourage the growth of more decentralized business and water service models.
- Given these technological developments, health and safety guidelines will need continuing consideration. This will require close cooperation across a number of areas of government policy. Ensuring proper oversight is required to underwrite sustainable outcomes.
- With water reuse now ubiquitous in many countries, the task of properly informing communities of contemporary risks attached to different reuse options and how they are managed needs to be an integral part of the policy framework and its implementation.
- Managing community perceptions around some supply augmentation options, in particular DPR, is critical. This can be achieved through structured consultation from the outset of option development and infrastructure planning. Political leadership will be

important.

- Even when technology is suitable and public acceptance strong, cost-effectiveness will be critical to ensure long-term acceptance and viability.
- Government regulators need to ensure they are using contemporary information in managing technology adoption, and have in place compliance regimes that can properly manage the risks from new technology.
- The mix of demand and supply options that make senses in an urban landscape will depend much on what currently exists, and the specific circumstances of the urban area in question. Technology surrounding DPR has reached a point where its introduction may well be justified in some circumstances. But before that can occur, the regulatory and compliance framework associated with this approach to water reuse is likely to need significant upgrading. This is likely to be a lengthy process.
- Appropriate data will assist in assessing the viability of competing options.
- Industrial reuse is driven by hard-nosed economic considerations, including brand impact. Often this may require a fundamental reassessment of water input. Detailed case studies provide a basis commencing this reassessment.

Acknowledgements

The author would like to thank several colleagues and the editor for useful comments on earlier drafts.

Disclosure statement

No potential conflict of interest was reported by the author.

References

Abrams, B., Kumaradevan, S., Sarafidis, V., & Spaninks, F. (2011). *The residential price elasticity of demand for water, joint research study*. Sydney. Retrieved from http://www.pc.gov.au/inquiries/completed/urban-water/submissions/sub083.pdf

ABS. (2006). 4610.0 – Water Account, Australia, 2004-05. Retrieved from http://www.abs.gov.au/AUSSTATS/abs@.nsf/allprimarymainfeatures/6F380840F971B08DCA2577E700158A5E?opendocument

ABS. (2014). 4610.0 – Water Account, Australia, 2012-13. Retrieved 13 March, 2014, from http://www.abs.gov.au/AUSSTATS/abs@.nsf/DetailsPage/4610.02012-13?OpenDocument

Aither. (2013). *Project report national validation framework for water treatment technologies—Summary information*. A report of a study funded by the Australian Water Recycling Centre of Excellence. Brisbane: Australian Water Recycling Centre of Excellence.

Albanese, A. (2006). Labor announces national water recycling target: Media Release Anthony Albanese MP [Press release]. Retrieved from http://anthonyalbanese.com.au/labor-announces-national-water-recycling-target

Aquacell Water Recycling. (2015). Aquacell case studies. Retrieved April 14, 2015, 2015, from http://aquacell.com.au/case-studies/

ATSE. (2013). Drinking water through recycling: The benefits and costs of supplying direct to the distribution system. A Report of a Study by the Australian Academy of Technological Sciences and Engineering (ATSE). Melbourne: Australian Academy of Technological Sciences and Engineering.

Australian Government. (n.d.). Policy and reform in the area of urban water. Retrieved 14 April, 2015, from http://www.environment.gov.au/water/cities-towns/policy-reform-urban-water

AWRCOE. (2014). Water recycling in food production and manufacturing summary report. Held at CSIRO Food and Nutrition, Werribee, Victoria, September 10, 2014.

AWRCOE. (2015). National Demonstration, Education & Engagement Program. Retrieved April 17, 2015, from http://www.australianwaterrecycling.com.au/projects/national-demonstration-education-amp-engagement-program

Bureau of Meteorology. (2015a). Climate resilient water sources. Retrieved July 31, 2015, from http://www.bom.gov.au/water/crews/site-explorer/

Bureau of Meteorology. (2015b). National water account. Retrieved April 15, 2015, from http://www.bom.gov.au/water/nwa/

Bureau of Meteorology. (2015c). *National performance report 2013–14: urban water utilities. Part A.* Melbourne: Bureau of Meteorology.

Byrnes, J. (2013). A short institutional and regulatory history of the Australian urban water sector. *Utilities Policy (2013), 24,* 11–19.

Chong, J. (2014). Climate-readiness, competition and sustainability: An analysis of the legal and regulatory frameworks for providing water services in Sydney. *Water Policy, 16,* 1–18.

Cook, S., Hall, M., & Gregory, A. (2012). *Energy use in the provision and consumption of urban water in Australia: An update. Prepared for the Water Services Association of Australia.* Australia: CSIRO Water for a Healthy Country Flagship.

Council of Australian Governments. (2004). *Intergovernmental agreement on the National Water Initiative.* Canberra. Retrieved from http://archive.coag.gov.au/coag_meeting_outcomes/2004-06-25/index.cfm-nwi

Council of Australian Governments. (2013). COAG councils. Retrieved April 15, 2015, from http://www.coag.gov.au/coag_councils

EPHC, NHMRC, & NRMMC. (2008). *Australian guidelines for water recycling: Augmentation of drinking water supplies.* Canberra: Environment Protection and Heritage Council, the National Health and Medical Research Council and the Natural Resource Management Ministerial Council.

Flow Systems. (2014). IPART review of Sydney Water Corporation's operating licence: Submission from Flow Systems Pty Ltd.

Flow Systems. (2015). Water services. Retrieved April 13, 2015, from http://flowsystems.com.au/water/water-services/

Frontier Economics. (2008). Approaches to urban water pricing (Waterlines Occasional Paper No 7). National Water Commission.

Fyfe, J., McKibbin, J., Mohr, S., Madden, B., Turner, A., & Ege, C. (2015). *Evaluation of the environmental effects of the WELS scheme, report prepared for the Australian Commonwealth Government Department of the Environment by the Institute for Sustainable Futures.* Sydney: University of Technology.

GHD. (2007). *Using recycled water for drinking: An Introduction Waterlines Occasional Paper No 2, June 2007.* Canberra: NWC.

Grafton, R. Q., & Ward, M. (2008). Prices versus rationing: Marshallian surplus and mandatory water restrictions. *Econ. Rec., 84,* S57–S65.

Gregory, A., & Hall, M. (2011). Urban water sustainability. In I. P. Prosser (Ed.), *Water* (pp. 75–88). Collingwood, VIC: CSIRO.

Grey, D., & Sadoff, C. W. (2007). Sink or swim? Water security for growth and development. *Water Policy, 9,* 545–571.

Guest, C. R. (2010). *Independent review of the water efficiency labelling and standards scheme.* Mimeo.

Harper, I., Anderson, P., McClusky, S., & O'Bryan, M. (2015). Competition policy review: Final report March 2015: Commonwealth of Australia.

Herndon, A. (2013, 1 May). *Energy makes up half of desalination plant costs: Study.* Bloomberg Business. Retrieved from http://www.bloomberg.com/news/articles/2013-05-01/energy-makes-up-half-of-desalination-plant-costs-study

Horne, J. (2013). Economic approaches to water management in Australia. *International Journal of Water Resources Development, 29*(4), 526–543.

Horne, J. (2015). Water information as a tool to enhance sustainable water management – The Australian experience. *Water, 7,* 2161–2183. doi:10.3390/w7052161

Humphries, K. (2014). Water Industry Competition Amendment (Review) Bill 2014, Second reading speech. Sydney: Retrieved from http://www.parliament.nsw.gov.au/prod/parlment/nswbills.nsf0/9D507AB3127DE421CA257D4E001E3F0B?Open&shownotes.

Hurlimann, A., & Dolnicar, S. (2010). When public opposition defeats alternative water projects – The case of Toowoomba Australia. *Water Research, 44*(1), 287–297.

Institute for Sustainable Futures. (2013). Looking to the future; Building industry capability to make recycled water investment decisions. Prepared by the Institute for Sustainable Futures, University of Technology Sydney, for the Australian Water Recycling Centre of Excellence.

IPART. (2014). *Review of the operating licence for Sydney Water Corporation: Water licencing-issues paper June 2014.* Retrieved from http://www.ipart.nsw.gov.au/Home/Industries/Water/Reviews/Licensing_-_Sydney_Water_Corporation/End_of_Term_Review_of_Sydney_Waters_Operating_Licence_2010_-_2015/16_Jun_2014_-_Issues_Paper/Issues_Paper_-_Review_of_the_Operating_Licence_for_Sydney_Water_Corporation_-_June_2014#thesubmission.

Khan, S. (2008). Urban reuse and desalination. In L. Crase (Ed.), *Water policy in Australia: The impact of change and uncertainty* (pp. 184–201). Washington, DC: RFP Press.

Law, I., & Hall, M. (2014). Resource recovery from wastewaters. Report on outcomes from the joint CSIRO/AWRCE Workshop, Melbourne, 7 March 2014. Synthesis Report 4 April 2014.

Marsden Jacob Associates. (2008). National snapshot of current and planned water recycling and reuse rates. Final report prepared for the Department of the Environment, Water, Heritage and the Arts June 2008. Melbourne: Marsden Jacob Associates.

Marsden Jacob Associates. (2013). Economic viability of recycled water schemes. A report of a study funded by the Australian Water Recycling Centre of Excellence. Brisbane: Australian Water Recycling Centre of Excellence.

Marsden Jacob Associates. (2014). Final project report working together to include potable water recycling in source development planning: A report of a study funded by the Australian Water Recycling Centre of Excellence. Brisbane: Australian Water Recycling Centre of Excellence.

Melbourne Water. (2013). Stormwater strategy: A Melbourne water strategy for managing rural and urban runoff. Melbourne: Melbourne Water.

Melbourne Water. (2014). Enhancing life and liveability: Melbourne water annual report 2013–14. Melbourne: Melbourne Water.

Melbourne Water Climate Change Study. (2005). Implications of potential climate change for Melbourne's water resources. A collaborative project between Melbourne Water and CSIRO Urban Water and Climate Impact Groups. Melbourne: Melbourne Water.

Milly, P. C. D., Betancourt, J., Falkenmark, M., Hirsch, R. M., Kundzewicz, Z. W., Lettenmaier, D. P., & Stouffer, R. J. (2008). Climate change e stationarity is dead: Whither water management? *Science, 319*(5863), 573–574. doi:10.1126/science.1151915

Moscovis, V. (2013). *Groundwater replenishment trial final report.* Perth: Water Corporation.

National Water Commission. (2014). *Urban water futures 2014.* Canberra: NWC.

NHMRC, & NRMMC. (2014). Australian drinking water guidelines (2011) – Updated December 2014. Retrieved March 9, 2015, from https://http://www.nhmrc.gov.au/guidelines-publications/eh52

NRMMC, EPHC, & NHMRC. (2009). *Australian guidelines for water recycling: Stormwater harvesting and reuse.* Canberra: Natural Resource Management Ministerial Council, the Environment Protection and Heritage Council, and the National Health and Medical Research Council.

NSW Government. (2014). Water Industry Competition Amendment (Review) Act 2014 No.57. Retrieved 17 March, 2015, from http://www5.austlii.edu.au/au/legis/nsw/num_act/wicaa2014n57516.pdf

NSW Office of Water. (2010). *Climate change and its impacts on water supply and demand in Sydney.* Retrieved from http://www.metrowater.nsw.gov.au/sites/default/files/publication-documents/climatechange_impact_watersupply_summary.pdf

Osmoflo. (n. d.). Wastewater treatment. Retrieved June 22, 2015, from http://www.osmoflo.com/en/our-capabilities/wastewater-treatment/

Paton, F. L., Dandy, G. C., & Maier, H. R. (2014). Integrated framework for assessing urban water supply security of systems with non-traditional sources under climate change. *Environmental Modelling & Software, 60,* 302–319.

Productivity Commission. (2011). *Australia's urban water sector: Final Inquiry Report (Vol. Report No. 55).* Canberra: Productivity Commission.

Quantum Market Research. (2014). *WELS Scheme Effectiveness Report of survey findings prepared for: Department of the Environment 11/07/2014. Job number: 14024.* Canberra: Quantum Market Research.

Quiggin, J. (2006). Urban water supply in Australia: The option of diverting water from irrigation. *Public Policy, 1*(1), 14–22.

Raucher, R. S., & Tchobanoglous, G. (2014). *The opportunities and economics of direct potable reuse.* Alexandria VA: WateReuse Research Foundation.

Samuel, G. (2014). Economic regulation, governance and efficiency in the Victorian water sector: Preliminary advice from the Independent Reviewer, May 2014. Retrieved from http://www.livingvictoria.vic.gov.au/PDFs/Fairer%20Water%20Bills/Preliminary%20Advice.pdf

seqwater. (2015). *Purified recycled water.* Retrieved 30 March, 2015, from http://www.seqwater.com.au/water-supply/water-treatment/purified-recycled-water

Short, M. D., Peirson, W. L., Peters, G. M., & Cox, R. J. (2012). Managing adaptation of urban water systems in a changing climate. *Water Resource Management, 26*, 1953–1981. doi:10.1007/s11269-012-0002-8

Sydney Water. (2014). *Water Efficiency Report 2013–14.* Sydney. Retrieved from https://www.sydneywater.com.au/web/groups/publicwebcontent/documents/document/zgrf/mdq3/~edisp/dd_047419.pdf

Water Corporation. (2014a). Fresh water thinking: Annual report 2014. Retrieved 16 April, 2015, from http://www.watercorporation.com.au/about-us/our-performance/annual-report/annual-report-2014

Water Corporation. (2014b). Groundwater replenishment. Retrieved March 20, 2015, from http://www.watercorporation.com.au/water-supply-and-services/solutions-to-perths-water-supply/groundwater-replenishment

Western Corridor Recycled Water Pty Ltd. (2009). Final progress report Western Corridor Recycled Water Project: Stage 2 - Scheme close out report. Retrieved from http://www.environment.gov.au/system/files/resources/91073add-a4e4-4942-b78c-7333e4fdbe76/files/wsa-qld-08-final-report.pdf

WSAA. (2006). Refilling the glass: Exploring the issues surrounding water recycling in Australia. *WSAA Position Paper, 2.*

WSAA. (2009). Meeting Australia's water challenges – Case studies in commercial and industrial water savings. *Occasional Paper No. 23.* Melbourne: Water Services Association of Australia Ltd.

WSAA. (2013). Seawater desalination information pack two. Melbourne: Water Services Association of Australia Ltd.

WSAA, & NWC. (2009). *National Performance Report 2007–08: Urban water utilities, March 2009.* Canberra: National Water Commission.

Wastewater reuse in Beijing: an evolving hybrid system

Olivia Jensen[a] and Xudong Yu[b]

[a]Lee Kuan Yew School of Public Policy, Institute of Water Policy, National University of Singapore; [b]Natural Resources Management, School of Environment and Natural Resources, Renmin University of China, Beijing

ABSTRACT

Water reuse capacity in Beijing has developed rapidly along hybrid lines, with a small number of large-scale plants connected to a network backbone and a large number of small-scale plants in less densely developed areas. This article examines whether Beijing's reuse system meets the objectives of effectiveness and sustainability, employing a new data-set of water reuse facilities for the city. It finds that reuse development in Beijing has so far been largely supply-driven and the desirable attributes of a hybrid system may only be achieved as greater attention is given to demand aspects of water reuse.

Introduction

The problems of water scarcity and water pollution in Beijing are well known (Xia, 2012). These twin pressures are strong motivation for investment in water reuse (Zhang et al., 2007), backed by the national policy framework, which has increasingly emphasized environmental quality (Shen, 2014). The expansion in water reuse capacity in Beijing in the last decade has been striking: the city government has set extremely ambitious targets for the collection and treatment of wastewater and the treatment of this wastewater up to levels suitable for certain types of reuse.

Scholars in the field of water engineering argue that hybrid systems, which combine large-scale treatment infrastructure and collection and supply backbones with small-scale distributed infrastructure and local networks, may be particularly robust, sustainable and efficient (Daigger & Crawford, 2007). Beijing's strategy for water reuse appears to exemplify a hybrid urban wastewater system. The city has a small number of large plants, mainly in the central urban area, and numerous smaller plants in the suburbs and rural areas of the municipality.

However, studies of water reuse in China suggest several potential concerns about the national approach to water reuse, which also apply to Beijing (Liu & Persson, 2013). First, in Beijing, as in other cities, there has been a clear focus on investment in treatment capacity but much less attention to constructing distribution networks so the reclaimed water can be used productively. Second, the government has sought to keep tariffs for reclaimed water low to stimulate demand, but it has not adequately addressed the issue of the financial sustainability of reuse investment projects and is filling the gap with subsidies, which may not be sustainable. Finally, standards lack consistency and completeness. Separate standards series for output water quality have been defined according to the end use of the reclaimed water, and standards for the storage, distribution and use of reclaimed water have yet to be set, raising the risks for potential reuse customers as well as the general public (Liu & Persson, 2013; Zhang et al., 2007).

The following section reviews the literature on hybrid urban water systems. This is followed by an overview of the development of Beijing's water reuse system, policy and management. The next section presents and analyzes data on the spatial configuration of Beijing's water reuse assets, and compares this to the configuration of a hybrid system. Plant-level data for more than 100 medium- and large-scale water reuse projects in Beijing which are currently operational or under development were collected for this analysis. Drawing on the data, the article offers some preliminary conclusions about how Beijing's framework for water reuse could be improved in order to deliver the desirable characteristics of an integrated hybrid system.

Hybrid urban water systems: a review of the literature

Hybrid urban water systems, which combine large-scale networks and treatment facilities serving large areas of the city with small-scale networks and treatment facilities, are thought to offer greater security and sustainability than other types of urban water systems such as a single plant–single network system or an entirely distributed system in which water and wastewater treatment are done at the community or household level (Daigger & Crawford, 2007). The advantages of these hybrid systems may include: better responsiveness to changing demand patterns; improved robustness to exogenous shocks and resilience to failures in parts of the system; lower net consumption for bulk water supply at the city level, thus freeing up resources for alternative non-urban uses; and lower energy requirements at the city level due to shorter conveyance and collection distances.

Water reuse can be integrated into three kinds of systems. In a unitary system (with a single plant or a small number of plants connected to a city-wide network), the wastewater treatment plant (WWTP) can be upgraded to provide recycled water for industrial, agricultural or potable (usually indirect) use. A separate distribution network is needed to deliver the reused water to customers. In a totally distributed system, water reuse can take place within the household or commercial building, with greywater being used for gardening, toilet flushing, etc. All these opportunities are available in a hybrid system.

Daigger and Crawford (2007) point out that many cities have these hybrid systems, although usually more by chance than by design. As cities have grown, new treatment facilities have been added, and as older areas of the city have been redeveloped, new technologies have been introduced through modification and upgrading of assets. Of course, it will be easier and cheaper to introduce some distributed treatment elements in a city that is consistently expanding compared to one which is undergoing partial and haphazard redevelopment.

While hybrid systems may seem appealing, they have several potential disadvantages related to their integration of small-scale systems: lower reliability associated with small plants; fewer economies of scale and scope, leading to higher costs; and governance issues associated with hybrid systems, notably the potential for poor management coordination between the different elements of the system, leading to gaps or overlap.

In the past, the main concern with distributed systems was the poor reliability of small-scale treatment facilities, so the risks to public health were considered too high for wide-spread use. This has changed in the last decade, mainly due to technological advances in small-scale membrane-based treatment processes like membrane bioreactors, which can reliably produce high-quality water on a small scale (Daigger, Rittmann, Adham, & Andreottola, 2005). Studies modelling health risks in different system structures have found that, *ceteris paribus*, health risks from waterborne diseases are lower in distributed systems, as a single source of contamination does not affect the whole distribution area (Fane, Ashbolt, & White, 2002). The cost of membrane-based systems has also fallen dramatically, bringing these technologies within reach of local governments.

Secondly, economies of scale are present for treatment plants and particular treatment processes, such as biosolids processing (Daigger & Crawford, 2007), militating for centralized systems. However, there are no economies of scale for networks, so whether a unitary or a distributed system is cheaper overall for a set volume of water delivered will depend on the comparative costs of plants and pipes. Fane et al. (2002) show that under certain reasonable assumptions, economies of scale peak at a system size of around 1000 connections and that above this level diseconomies of scale are present.

Thirdly, while coordination is certainly desirable to ensure complementarities and avoid overlap between the different parts of the urban water system, it does not necessarily require all elements to be under unified management. Conversely, even when a single organization is responsible for all water and wastewater services, vertical divisions within the organizational structure can lead to the same result.

In this discussion of unitary, distributed and hybrid urban water systems, it is important to note that this categorization relates to the city's physical infrastructure, not the policy or planning process. It is possible, and indeed quite likely, that an efficient and sustainable hybrid system would be the result of a single, integrated plan for the whole of the urban area. To distinguish between these two, the terms 'centralized' and 'decentralized' may be used to refer to the spatial level at which policy and planning takes place; 'integrated' and 'segregated' to refer to the extent to which policy and planning are coordinated across functional areas (water resources, water supply, stormwater, wastewater, water reuse, etc.); and 'unitary' and 'distributed' for the size and type of physical infrastructure.

Water reuse in Beijing

Policy and planning

In China, informal reuse of wastewater for agricultural purposes was common historically, but the practice of sewage irrigation increased rapidly in the 1980s and 1990s as economic reforms were implemented (Liu & Persson, 2013). Policy and technology developed further during the 6th Five-Year Plan (1981–1985), during which water reuse was included as a special project under the National Special Science and Technology Plan and the first pilots were carried out in the cities of Dalian and Qingdao. The government declared that urban wastewater could be reused as a kind of water resource after being treated by the municipal water reuse plant (WRP) in Dalian and Qingdao (Yi, Jiao, Chen, & Chen, 2011). A national policy for water reuse was launched and city-level planning targets were defined in the early 2000s.

Table 1. Water reuse policy in Beijing.

Year	Policy document	Key content
1987	Tentative Management Measures for Reclaimed Water Infrastructure Construction	Basic rules for water reuse
1991	Urban Water Conservation Regulations of Beijing	Binding regulations on effluent quality
1994	Sanctions for Waste Water Discharge in Beijing	Penalties and corrective measures for plants and building not conforming to 1987 measures
2001	Notices about Strengthening the Construction of Water Facilities Management	Requirement for all properties with construction area greater than 50,000 m² to build wastewater treatment facilities
2002	2008 Beijing Olympics Action Plan	Set a target rate for urban wastewater treatment of 90% and a rate of treatment to reuse quality of wastewater treatment plant effluent water of 50%
2003	City Water Price Adjustment – Beijing Municipal Price Bureau	Set the price of reclaimed water at RMB 1/m³; exempted reuse water users from sewage disposal fee
2003	Emergency Management Measures about Prevention and Control of SARS in Urban and Rural Community Construction and Public Environment Space – Ministry of Construction	Suspended the use of reclaimed water in the isolation zone and in residential areas
2004	Issues on Forming the Beijing Water Authority – General Office of the Beijing Government	Beijing Water Authority established four-level water management system: municipal; district/county; grass-roots water station; water users association
2005	Beijing Water Saving Method	Responsibility for the management of water reuse conferred to Beijing Water Authority
2006	11th Five-Year Plan (2006–2010)	Strategic Plan for the Utilization of Reclaimed Water
2009	Beijing Drainage and Reclaimed Water Management Methods	Clarification of the water reuse management system
2011	Beijing's Water Resources Protection and Utilization Plan for the 12th Five-Year Plan (2011–2015) – Beijing Water Authority	Set targets for the expansion of capacity and utilization, make steps for annual plan and regional plan
2012	Opinions on Further Strengthening Sewage and Reclaimed Water Utilization	Set main goals for 2012–2015
2013	Action Plan to Accelerate the Construction of Sewage Treatment and Recycled Water Facilities (2013–2015)	Outlined responsibilities and allocated specific tasks to different bureaus

Source: Beijing Municipal Government; Beijing Water Authority; Beijing Development and Reform Commission; Beijing Municipal Administration Commission.

Although Beijing was not the first city to engage in water reuse, it has developed quickly, driven by concerns over water scarcity and the related over-exploitation of surface and groundwater. The government first introduced guidance for the construction of WRPs in 1987, after which reclaimed water was used in some large buildings and experimental plots. However, after the SARS outbreak in 2003 led to heightened perceptions of public health risks, reuse was halted for a short period.

Winning the bid for the Beijing Olympics of 2008 gave new impetus to improving environmental performance, including wastewater treatment and reuse. Beijing was already ahead of other Chinese cities in its regulation of wastewater treatment and treatment standards, and these were tightened further as the city prepared for the Olympics. In the 12th Five-Year Plan (2011–2015), the government reiterated its commitment to the development of water reuse, primarily driven by resource scarcity, and set ambitious targets for the expansion of reuse capacity. Relevant policies and targets are listed in Table 1.

As a result of these proactive policies, the contribution of reclaimed water to total water supply has grown from a negligible level in 2002 to 22% in 2014. This is a remarkable

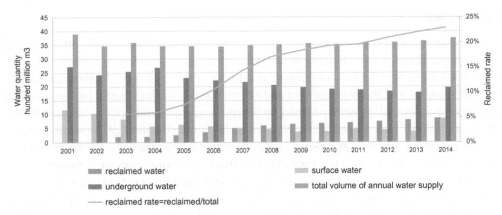

Figure 1. Water supply sources in Beijing Municipality, 2001–2014. Source: Beijing Water Authority.

Table 2. Target plans of Beijing wastewater treatment and reclaimed water reuse (2015).

Target item	Inner city	New city	Towns	Villages	Beijing
Upgrade and expansion of wastewater treatment plants	5	12	8	250*	25
Increased wastewater treatment capacity (10,000 m³/d)	134	76.6	17.7		228.3
Wastewater treatment ratio	100%	90%		60%	90%
Increased sludge disposal capacity (t/d)	1,800	2,195			3,995
New sewage pipeline (km)	348	612	249		1,209
Upgrade sewage pipeline (km)	353	600	443		2,116
New reclaimed water pipeline (km)	158	326			685
Increased water reuse plants	11	18	21		50
Increased water reuse plant capacity (10,000 m³/d)	350	144.6	18.3		513
Reclaimed water reuse ratio					75%
Increased reclaimed water reuse capacity (m³/y)					≥ 1 billion

*Small wastewater treatment station.
Source: Beijing Municipal Government, 'Opinions on Further Strengthening Sewage and Reclaimed Water Utilization' (2012) and '2013–2015 Beijing Water Reuse Plan' (2013).

expansion by any account, as shown in Figure 1. The quantity of reclaimed water reached 680 MCM (million cubic metres) in 2010, more than the quantity of surface water used in Beijing. This increase appears to have allowed a reduction in the exploitation of groundwater, from over 70% of the water supply in 2003 to 52% in 2014.

The ambitious targets for the development of water reuse capacity under the 12th Five-Year Plan include an overall target for water reuse capacity of 4.6 MCM/day by the end of 2015. Under the plan, several different types of projects are envisioned in an integrated approach, including wastewater treatment, water reuse, pipe network construction and the safe disposal of sludge. The detailed targets are shown in Table 2.

The plan is centralized, in that it is set by the Beijing government for the entire municipality; and it follows hybrid principles: different targets are set for the administrative subdivisions within the municipal boundary according to their level of economic and social development. Sewage treatment facilities are required to be built in towns and villages, although their capacity is quite small. The inner city has higher wastewater treatment and water reuse targets by proportion and volume: there, the wastewater treatment ratio should be 100%, and WRP capacity should reach 3.5 MCM/day by the end of 2015.

Governance

In most Chinese cities, water supply, wastewater, water reuse and drainage fall under different government bureaus, including the water resources/water conservancy bureau, the environmental protection bureau, the construction bureau (the housing and rural-urban development bureau), and the development and reform bureau. The local water and wastewater utility companies, usually wholly owned by the municipal government, are responsible for implementation (Cosier & Shen, 2009). Water reuse falls under the remit of the water conservancy department, which is responsible for developing 'unconventional' water sources, while the construction department is responsible for urban water supply infrastructure planning, and the companies for construction and operation. This complex structure makes integrated planning of the water sector difficult. Departments responsible for policies relating to potential users of reclaimed water, like industry and irrigated agriculture, are not represented in the policy-making process for water reuse.

Beijing has sought to improve coordination between these different aspects of water planning – water resources, water supply, wastewater and reuse – by creating an integrated governance structure with a single administrative agency, the Beijing Water Authority (BWA). BWA plays a role in collecting and analyzing data on water supply, helps coordinate the planning process, and designs and procures some projects. Although BWA does not have any official administrative status that would allow it to enforce targets or recommendations on other government agencies, it plays a critical role in coordinating investment planning. The establishment of BWA made possible the coordination of the demand and supply aspects of water reuse, identification of customers, and location of infrastructure investment.

Several public utilities are involved in water reuse in Beijing: the Beijing Waterworks Group and the Beijing Drainage Group both have water reuse subsidiaries. In addition, two exchange-listed water companies, majority-owned by the municipal government, Beijing Capital and Beijing Enterprises Water Group, are active in the sector as investors and contractors. Finally, several fully private companies are engaged in water reuse projects under the public–private partnership (PPP) model, such as Beijing Origin Water. Water affairs bureaus at the district level (the level of government below the Beijing Municipal Government) have been given the authority to design and award these PPP contracts and have actively taken up the model as a way of meeting demanding targets for upgrading wastewater treatment requirements within strict budget constraints.

In addition to private actors involved in the system through PPPs, institutional and commercial developers and facilities managers also play a role in Beijing's complex system. In 1987, the municipal government brought in regulations requiring all new buildings with a construction area greater than 30,000 m^3 to install their own wastewater reuse systems. The standards for the wastewater source, reuse technology and discharge water were further detailed in regulations issued in 2000. It is difficult to assess the contribution of these plants to total capacity, as data on the number and size of these plants are sparse. One study on small-scale systems identifies a range of capacity of 65–400 m^3/day and estimates that there were 1000 such plants in the city in 2009 (Liang & van Dijk, 2010). Assuming an average size of 200 m^3/day, this would amount to an extra 3% water reuse capacity in the city on top of the municipal plants. Given the lack of reliable data on these private facilities, the data-set used for this article covers only municipal WRPs.

In Beijing, WRPs are responsible for monitoring their own water discharge quality. Industrial users are expected to monitor the quality of recycled water they receive, according to industrial water quality standards (GB/T 19923-2005). Users are responsible for maintaining sterilization algae, stable water quality, the yield of water, and water equipment monitoring and control. There are also third-party monitoring agencies, such as the Beijing Urban Drainage Monitoring Centre Company, which falls under Beijing's State-Owned Assets Management Company. This is a professional institution monitoring drainage water quality and water quantity in six Beijing districts. Since 2003, it has also monitored reclaimed water quality.

Use of reclaimed water

The expansion of reclaimed water supply has occurred alongside a rebalancing in the consumption of water from industry use towards environmental use. In 2003–2014, the proportion of water allocated to environmental rose from under 1% to 19%. These shifts are consistent with the higher priority assigned to environmental quality in Beijing policy.

There are also some instances in Beijing of the use of reclaimed water for power plant cooling and as process water in manufacturing, and a small amount of high-quality reclaimed water is reused as a raw material in production as well as for forestry irrigation, suggesting that the barriers to using reclaimed water in productive sectors are not insuperable. For example, the BOE Technology Group Co., an electronics company in Beijing, uses reclaimed water in the production of liquid crystal displays. Reclaimed water accounts for more than 90% of its total water consumption. The company is in the Beijing Economic and Technological Development Area, an industrial zone, where a WRP and reclaimed water supply network were built into zone development plans. Reclaimed water there is supplied at RMB 5.5/m^3, compared to RMB 6.21/m^3 for ordinary industrial water.

Micro-analysis of Beijing's reuse system structure

In order to analyze the hybrid nature of Beijing's water reuse system in more detail, a dataset of medium- and large-scale water reuse projects was constructed. Data were drawn from official sources, principally the Beijing Municipal Government, BWA, company reports and Internet searches, verified through interviews. Small-scale water reuse facilities with a capacity of under 500 m^3/day treating water from individual institutional or commercial premises and equipment for recycling water within industrial facilities are not included in the data-set. In the data-set, each upgrade or extension of a facility is classified as a separate 'project' even if it shares the same location and influent stream, in order to track the development of reuse capacity over time. Some 118 projects which are either operational, under construction or planned were identified.

Table 3 shows clearly the rapid growth of water reuse capacity in Beijing in the 15 years beginning in 2000. However, this rate of development will be vastly surpassed if the planned expansion in 2015 and 2016 is realized. In these two years, the total water reuse capacity of the projects in the data-set will increase by about 200% if all planned projects are completed. While it is possible that these facilities may not be completed and commissioned within that short time period, it seems likely that the projects will be realized in the next 3–5 years as contracts have already been awarded and in many cases the plants are already under construction.

Table 3. Water Reuse Capacity by Year and District Type (m³/day).

District type	2000	2003	2004	2005	2006	2008	2009	2010	2011	2013	2014	2015	2016	Total
Rural	20,000				550,000	80,000	55,000	123,000	20,000	357,700	544,200	634,500	305,200	2,039,600
Urban		40,000	105,000				20,000	80,000		85,000	20,000	1,870,000	1,500,000	4,370,000
Total	20,000	40,000	105,000		550,000	80,000	75,000	203,000	20,000	442,700	564,200	2,504,500	1,805,200	6,409,600
Cum. total	20,000	60,000	165,000	165,000	715,000	795,000	870,000	1,073,000	1,093,000	1,535,700	2,099,900	4,604,400	6,409,600	

Source: Compiled by authors from data in the *Beijing Water Resources Bulletin*.

Table 4. Water Reuse Capacity and Economic Indicators, by District.

District	District type	Total water reuse design capacity (m³/day)	Largest project reuse design capacity (m³/day)	Smallest project reuse design capacity (m³/day)	No. of projects	First project operational (year)	Population, 2013 (thousands)	Industrial output, 2013 (million RMB)	Irrigated area, 2013 (km²)	Capacity percapita	Capacity/industrial output	Capacity/irrigated area
Changping	Rural	372,800	80,000	2,000	20	2009	1889	127,099	57.4	197.35	2.93	6,494
Chaoyang	Urban	2,320,000	600,000	20,000	14	2000	3,841	109,315	24.6	604.01	21.22	94,329
Daxing	Rural	375,900	120,000	3,000	10	2010	1,507	63,311	352.9	249.44	5.94	1,065
Dongcheng	Urban	0	0	0	0		909	12,355				
Fangshan	Rural	134,500	60,000	1,200	10	2013	1,010	97,073	192.4	133.17	1.39	699
Fengtai	Urban	1,400,000	600,000	5,000	8	2003	2,261	41,862	12.4	619.20	33.44	112,582
Haidian	Urban	800,000	550,000	10,000	6	2006	3,576	170,831	9.3	223.71	4.68	85,714
Huairou	Rural	265,000	130,000	35,000	4	2009	382	55,322	66.6	693.72	4.79	3,980
Mengtougou	Rural	160,000	80,000	40,000	4	2010	303	8,071	0.8	528.05	19.82	190,931
Miyun	Rural	85,000	65,000	20,000	3	2014	476	29,306	57.1	178.57	2.90	1,488
Pinggu	Rural	178,000	80,000	7,000	9	2010	422	24,259	96.6	421.80	7.34	1,842
Shijingshan	Urban	20,000	20,000	20,000	1	2013	644	27,655		31.06	0.72	
Shunyi	Rural	216,900	80,000	3,000	15	2014	983	286,253	285.8	220.65	0.76	759
Tongzhou	Rural	253,300	160,000	13,300	5	2014	1,326	66,772	270.8	191.03	3.79	935
Xicheng	Urban	0	0	0	0		1,303	88,319				
Yanqing	Rural	63,700	40,000	2,600	9	2015	316	6,503	183.4	201.58	9.80	347

Source: Beijing Statistical Yearbooks, Beijing Water Authority.

Table 4 shows the geographical distribution of water reuse facilities. Until 2013, water reuse was concentrated in the central urban districts of Beijing municipality. Since then, there has been a dramatic increase in investment in water reuse in the outer districts, which are classified as rural (these outer districts have mixed land-use patterns, with residential, commercial and industrial areas in addition to agriculture and forestry). This acceleration of investment in outlying areas appears to have been driven by targets set out in the city's five-year plan for wastewater treatment rather than by demand for reclaimed water in these locations.

The first reuse projects in Beijing were in the populous inner urban districts of Chaoyang and Fengtai, adjacent to two of the city's largest wastewater treatment facilities. This is also reflected in the size of the reuse projects in these locations: in both districts, the largest project has a capacity of 600,000 m³/day.

Not surprisingly, the smallest reuse plants are located in the outer districts. The smallest project recorded in the database is a 1200 m³/day plant in Fangshan. The inner urban areas of Dongcheng and Xicheng are the most densely built parts of the city and have no large wastewater treatment facilities. Wastewater is discharged and piped out to surrounding districts for treatment, reflecting space constraints and concerns about odour in the densely developed central urban areas.

The geographical pattern of reuse in Beijing is consistent with the hybrid urban water system model, including both centralized reuse facilities located next to main points of supply – the major WWTPs – and many smaller facilities located in outlying areas. But to what extent does this pattern of investment reflect the distribution of demand in the city? Figures 2, 3 and 4 show the relationship between reuse capacity by district and three district-level indicators: population, industrial output and irrigated land area, respectively. Population is taken as a proxy for supply of input water for reuse, while industrial output and irrigated land area are taken as proxy measures of demand. Industry and agriculture are considered the main potential productive uses of reclaimed water in China as reclaimed water is not yet supplied for potable use (either direct or indirect).

The data support the view that the development of reuse capacity has been supply-driven. They show a strong positive relationship between population and WRP capacity (Figure 2), a weak positive relationship between industrial output (heavy industry) and WRP capacity

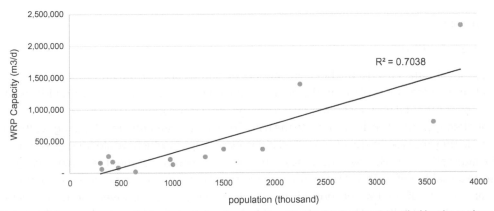

Figure 2. Water reuse plant design capacity and population, by district. Source: Compiled by the authors; population data from Beijing Municipal Government.

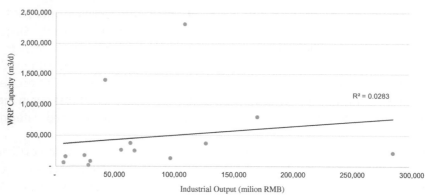

Figure 3. Water reuse plant design capacity and heavy industrial output, by district. Source: Compiled by the authors; output data from Beijing Municipal Government.

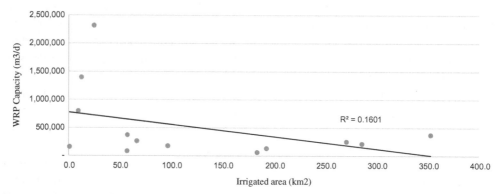

Figure 4. Water reuse plant design capacity and irrigated area, by district. Source: Compiled by the authors; irrigated area data from Beijing Municipal Government.

(Figure 3), and a negative relationship between the land area of irrigated agriculture and WRP capacity (Figure 4). Note that the data for these figures include projects planned for completion in 2015 and 2016 where data on capacity were available, so the relationships shown in the figures are not simply due to the fact that reuse investment in the outer areas is more recent and can be expected to catch up soon.

Discussion

Uses and utilization

Plant-level data provide further evidence for the emphasis on environmental uses for reclaimed water: most plants in the database, including those in rural areas, provide water for environmental uses, including river recharge and landscaping (Table 5). These uses are not directly productive (in contrast to the use of reclaimed water for industry or agriculture), nor do they replace the use of municipal piped water supply, which would free up resources for alternative uses. However, they may enhance the urban environment and therefore generate social benefits. This tallies with national trends: the main use of reclaimed water in China (45.4% of total water reuse) is for 'scenic environment' use (river and urban water body recharge), followed by industrial use (23.7%) and irrigation for agriculture and forestry (23.4%; data for 2009 from Zhong et al., 2012, quoted in Liu & Persson, 2013).

Table 5. Water Reuse Application by Water Reuse Plants.

District	District type	Water reuse design capacity (m³/day)	Application
Changping	Rural	30,000	Landscape water, urban miscellaneous
Changping	Rural	60,000	Landscape water, urban miscellaneous
Changping	Rural	20,000	Mengzuhe River recharge
Changping	Rural	80,000	Landscape water, urban miscellaneous, industrial water, wash water, power plant cooling water
Changping	Rural	70,000	30,000 m³/day landscape water, urban miscellaneous; 40,000 m³/day Lingou River recharge
Changping	Rural	20,000	Dongshahe River recharge, urban miscellaneous
Changping	Rural	17,700	Landscape water, forestry irrigation
Chaoyang	Urban	200,000	Landscape water
Chaoyang	Urban	300,000	Nandagou River recharge; future phases will supply 150,000 m³/day landscaping, urban miscellaneous, industrial cooling water
Chaoyang	Urban	20,000	Irrigation
Chaoyang	Urban	100,000	60,000 m³/day power plant cooling; 20,000 m³/day irrigation, urban miscellaneous; 20,000 m³/day Bahe River recharge
Chaoyang	Urban	40,000	Emissions into Qinghe River, which became the important water source for the Olympic park
Chaoyang	Urban	60,000	Landscape water (Olympic park); 10,000 m³/day toilet flushing, cleaning
Chaoyang	Urban	600,000	70,000 m³/day industrial use; landscape water
Chaoyang	Urban	47,000	200,000 m³/day transfer to Gaobeidian Lake as power plant cooling water; 100,000 m³/day industrial use; urban miscellaneous
Chaoyang	Urban	20,000	Irrigation, spraying roads, landscape water, toilet flushing
Daxing	Rural	120,000	Landscape water
Daxing	Rural	40,000	Landscape water
Daxing	Rural	40,000	Xihongmenzhen River recharge, landscape water
Daxing	Rural	3,000	Tiantanghe River recharge
Daxing	Rural	20,000	Industrial resue water (electronics industry, etc.)
Daxing	Rural	20,000	Industrial resue water (electronics industry, etc.)
Daxing	Rural	100,000	River recharge
Fengtai	Urban	600,000	Landscape water
Haidian	Urban	80,000	Part pipelined into communities and used for flushing or green irrigation; other part discharged into the downstream of the Nansha River as channel-filling water
Haidian	Urban	20,000	Youyiquhe River recharge
Haidian	Urban	20,000	Dongfutougou River recharge
Mentougou	Rural	40,000	Landscape water, urban miscellaneous
Mentougou	Rural	80,000	15,000 m³/day landscape water; 65,000 m³/day urban miscellaneous
Pinggu	Rural	40,000	Landscape water, urban miscellaneous, industrial water, irrigation water
Shijingshan	Urban	20,000	Gaojingou River and Yongdinghe River recharge
Huairou	Rural	35,000	Huaihe River recharge
Huairou	Rural	100,000	Huaihe River recharge

Source: Compiled by the authors.

The potential for reclaimed water to replace piped water in Beijing would appear to be large but in most places has been held back, possibly by the absence of a distribution network. Such a network is very costly to construct, especially in the densely developed urban areas (Zhang et al., 2007), so reclaimed water in Beijing is delivered to most customers via tanker (with the exception of large industrial users close to WRPs). This greatly increases overall costs of supply (Liang & van Dijk, 2010).

Another possible use for reclaimed water in China is groundwater recharge. In Beijing, as in other cities in China, groundwater has been over-exploited, raising risks of land subsidence, and has become highly contaminated, leading to public health risks. The social benefit of this type of water reuse would probably be high in China. Although the technologies exist and using reused water for groundwater recharge is widespread in the US and other countries, it has not so far been practised in Beijing or elsewhere in China. This is mainly because of concerns over quality – groundwater recharge requires high, consistent levels of water quality – and cost.

While there is no doubt that reuse capacity in China has grown rapidly, a concern with focusing on WRP capacity is whether this capacity is actually used. Consistent and complete utilization data are not available from government data sources. Aggregate data on water supply sources are available in the *Beijing Water Resources Bulletin*. However, the volume of reclaimed water supplied includes water supplied by non-municipal plants and small plants that are not included in the database. As data on the capacity of these plants are not available separately, these data cannot be used to estimate the overall utilization rate of WRPs in the city.

Given these limitations in existing data, data on utilization were collected, where available, at the plant level. Data on utilization were included for 25 of the plants in the database. These had a range of 12–118%, with an average of 79%. (Utilization rates above 100% are possible if the plant is run above its design capacity.) This high level of utilization in some plants, and the broad spread in rates between plants, suggest that the fundamental challenge in Beijing is not a lack of demand for reclaimed water but the need to match demand and supply in specific localities better.

Data on utilization rates elsewhere in the country are very limited. One point of comparison is a study of nine WRPs in China in 2005, which found an average utilization rate of less than 50% and a lowest value of 10% (Zhang et al., 2007), suggesting that utilization rates in Beijing are comparatively high.

Quality

One of the constraints on the use of reclaimed water for productive uses, or ultimately for indirect potable use, is inadequate quality. The confusing web of standards that are applied to WRPs in China may make potential customers more cautious of switching to reclaimed water.

On the one hand, standards imposed on discharges from WWTPs are specific to the type of plant and technology used, distinguishing between traditional and membrane technologies. According to the standards, wastewater treated to standard 1A is the basic requirement for reclaimed water. However, the discharge water quality of most treatment plants does not reach 1A, and water is discharged to the environment or reused at standard 2. Only 17 of the WWTPs in the data-set meet the highest treatment standard, which makes water suitable for all non-potable uses without further treatment (Table 6). A few WRPs produce high-quality reclaimed water above the 1A level, but there is currently no water quality standard to fit this case.

The design of urban WWTPs is based on discharge standards, not on water reuse, so there is a great difference in the standards necessary for certain key parameters like nitrogen and phosphorus. The Beijing government is seeking to improve the quality of wastewater

Table 6. Water Reuse Plant (WRP) Discharge Standards.

Project	Wastewater discharge standard	Project	Wastewater discharge standard
Changping Weilai Technical Town WRP	1B	Mentougou WRP	1A
Changping Xiaotangshanzhen WRP phase 1	1B	Mentougou 2nd WRP	1A
Chaoyang Qinghe 2nd WRP phase 1	1B	Mentougou Mencheng WWTP	2
Chaoyang Qinghe 2nd WRP phase 2	1B	Pinggu WRP	1A
Chaoyang Qinghe 2nd WRP phase 3	1B	Shijinshan Wulituo WWTP	1A
Chaoyang Dingfuzhuang WRP	1B	Shunyi Beiwuzhen WRP	1A
Chaoyang Fatou WRP	1B	Shunyi WWTP upgrade WRP	1B
Chaoyang Gao'antun WRP	1B	Yanqing Zhangshanyin zhen WRP	1A
Chaoyang Xiaohongmen WWTP upgrade to WRP	2	Yanqing Dayushuzhen WWTP	1A
Daxing Huangcun WWTP upgrade to WRP	1B	Yanqing Xiangyingxiang WWTP	1A
Daxing Tiantanghe WWTP upgrade to WRP	1B	Yanqing Jingzhuangzhen WWTP	1A
Daxing Xihongmen WRP	1B	Yanqing Kangzhuang WWTP	1A
Fangshan Zhoukoudian WWTP	1B	Yanqing Badalingzhen WWTP	1A
Fangshan Chenguan WRP	1B	Yanqing Yongningzhen WWTP	1A
Fangshan Liangxiang WRP	1A	Yanqing Jiuxianzhen WWTP	1A
Fengtai Hexi WRP	2	Yanqing County Chengxi WRP	1A
Fengtai Qinglonghu WRP	1B	Huairou WRP phase 1	1A
Haidian Daoxianghu WRP phase 1	1B	Huairou WRP phase 3	1A
Haidian Yongfeng WRP phase 1	1B	Haidian Cuihu WRP	1B
Haidian Wenquan WRP phase 1	1B	Haidian Shangzhuan WRP phase 1	1B
Haidian Qinghe WRP	2		

Source: Compiled by the authors.

Note. WWTP = wastewater treatment plant. The 1A standard is the basic requirement for urban WWTP effluent that will be reused. WWTPs discharging effluent into smaller lakes as landscape water or general reuse water must meet the 1A standard. Urban WWTP effluent discharged into surface water at class III in GB3838 (except designated drinking water sources areas and swimming areas), or seawater at class II in GB3097, or closed or semi-closed waters such as lakes and reservoirs, must meet the 1B standard. Urban WWTP effluent discharged into surface water at class IV or V in GB3838, or seawater at class III or IV in GB3097, must meet the second standard. WWTPs using first-grade strengthening treatment technology, which are not in key control basins or water source areas, must meet the third standard according to local economic conditions and water pollution control requirements. And a location must be reserved for secondary treatment facilities, to reach the secondary standard. Discharge standard of pollutants for municipal wastewater treatment plant GB18918-2002.

discharges and to connect distribution networks to achieve an integrated reclaimed water pipe network consistently supplying 1A-standard water.

At present, there are several national standard series for reclaimed water quality, differentiated by the intended use of the recycled water. The categorization was set out in 2002, and detailed standards were subsequently issued for urban use (2002), scenic environment use (2002), groundwater recharge (2005), industrial use (2005), agricultural irrigation (2007)

Table 7. Wastewater reuse technologies.

Technology	No. of projects	Technology	No. of projects
3AMBR	3	MBR + ozone disinfection	1
A2/O	8	MF/UF-RO	2
A2/O + depth filtration	4	MSBR + ABF	1
A2/O + MBR	6	Oxidation ditch	6
A2/O + UM	2	Oxidation ditch + UM	1
Activated sludge	2	SBR	5
Cyclic activated sludge system	1	SBR + MBR	1
MBR	13		
Total	**56**		

Note: 3AMBR = anoxic-anaerobic-anoxic membrane bio-reactor. MBR = membrane bio-reactor. 3AMBR = anoxic -anaerobic-anoxic membrane bio-reactor. A2/O = anaeroxic-anoxic-oxic process. SBR = sequencing batch reactor. UF = ultra filtration. MF = micro filtration. RO = reverse osmosis.
Source: Compiled by the authors.

and green space irrigation (2010). In addition, any specific use of reclaimed water should meet the industry standard for water supply as for tap or groundwater. Comparing these quality standards with the 1A wastewater discharge quality standard shows that some key parameters are stricter, while others may not be as demanding as the 1A standard.

From the outset, Beijing has maintained some of the strictest wastewater discharge standards in China. For example, the Beijing Water Pollutant Discharge Standard (trial) of 1985 was stricter than the national-level Integrated Wastewater Discharge Standard (GB8978-88). Beijing municipal standards are being gradually tightened and made more comprehensive.

The data-set includes information on technologies used in 56 projects (Table 7). Of these, 13 use membrane bioreactors (MBR) without any additional treatment, and a further 11 use MBR in combination with another form of treatment, including one that uses disinfection. MBR is capable of delivering consistently low levels of suspended solids in output water, making disinfection more effective, and is considered a technology well adapted to distributed reuse systems. However, this depends on systems being operated correctly and on the regular maintenance and replacement of membranes to keep up treatment performance.

Environmental performance

Given that the primary use of reclaimed water is for river recharge, it is interesting to consider whether the quality of water in Beijing's rivers and lakes has improved. Data from the Ministry of Environmental Protection (Figure 5) show that quality improved during the years leading up to the 2008 Beijing Olympic Games, but there has been an apparent decline since 2011. The quality of water in reservoirs has also declined. Given the significant increase in wastewater treatment and water reuse capacity in this period, there are several possible explanations for the decline in environmental quality. One possibility is that plants are operating above capacity and so are not able to meet design standards. Another possible factor is an increase in diffuse sources of water pollution over the period. Additionally, the recorded concentration of pollutants will be higher if the quantity of surface water declines. While recycled water may contribute to both quantity and quality of surface water, it may be not be sufficient to compensate for abstractions and pollutant discharges. This suggests that the use of reclaimed water for river recharge must be accompanied by a host of other regulations and measures if it is to deliver improvements in environmental outcomes.

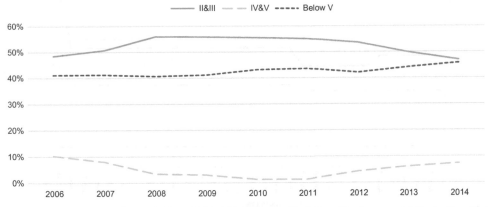

Figure 5. River water quality in Beijing, 2006–2014. Source: Ministry of Environmental Protection (data downloaded May 2015). Note: According to The Surface Water Environmental Quality Standard of China (GB 3838-2002), I - class is mainly in the source water and state reserve; II - class is mainly 1-level centralized domestic drinking water protection zones, rare aquatic habitats, fish spawning grounds, larvae of larval feeding, etc; III - class is 2-level centralized domestic drinking water protection zones, shrimp, wintering grounds and migration channels, aquaculture and fishery waters and swimming; IV - class is general industrial water and entertainment water which is not direct contact with body; V- class is mainly agricultural water and landscape water; Below V - class is water that doesn't reach the quality standard of class V.

Financial sustainability

Prices for recycled water are set at the municipal level. In Beijing, the standard price of recycled water is RMB $1/m^3$, below the price for regular piped supply, to stimulate demand. Tariffs for reclaimed water have not been raised since they were first introduced in 2003. Reclaimed water is also exempt from water resource charges and sewage treatment charges, which are levied on top of the base price for piped water supply. Over this period, the tariff for tapwater has increased, to RMB 5 in 2014, widening the margin with the reclaimed water tariff and therefore strengthening the incentives of customers to switch to reclaimed water. The evolution of tariffs is shown in Figure 6. However, the low level of take-up of reclaimed water by industry suggests that factors other than price may be influencing consumers' decisions.

The suppression of tariffs also undermines the financial sustainability of water reuse projects. The plant location, technology, scale, water quality and utilization of capacity all affect the cost of treatment. Tan et al. (2015) estimate that the total cost (capital and operating cost) is on average RMB $5.46/m^3$ for a WWTP, using data from a sample of 227 plants in China. Further cost data are provided by Yang and colleagues based on a sample of WRP projects across China. They find operating costs of RMB $0.4–1.5/m^3$ and total costs of RMB $0.56–3.0/m^3$ (Yang, Ren, & Hu, 2011, quoted in Liu & Persson, 2013), with total costs above RMB 1 in half the projects in the sample. Specifically for Beijing, the authors found operating costs of WRPs under the build-operate-transfer model of RMB $1.7/m^3$, of which RMB $0.7/m^3$ is covered by a subsidy from the government. It is important to note that these cost estimates do not include capital or operating costs for standard wastewater treatment or distribution. Furthermore, costs will be higher for plants using more sophisticated technologies. Operating costs for a membrane WRP in Beijing are likely to be around RMB $2/m^3$.

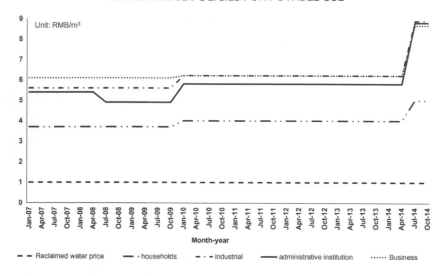

Figure 6. Water tariffs in Beijing, by source and customer category, 2007–2014. Source: Beijing Water Authority.

This leaves a significant gap between the costs and tariffs which needs to be covered by a government subsidy. Assuming an operating cost of RMB 2/m³ plus an additional 25% to reflect capital costs, 20% to reflect the costs of traditional treatment and 40% to reflect the costs of constructing and operating the distribution network, a tariff of RMB 3.7/m³ would be needed to cover costs. At the 2014 level of reclaimed water production of 860 MCM, the total subsidy cost to government would amount to RMB 2.3 billion annually. While it may be possible for the municipal government to provide this subsidy in the short term to promote the development of the sector, in the longer term tariffs will need to be increased to allow the recovery of costs. To preserve incentives for customers to use reclaimed water over tapwater, tapwater tariffs could be increased further.

Notwithstanding the difficulty of structuring WRPs in a financially sustainable manner, district governments in Beijing have made extensive use of PPP models to finance infrastructure investment upfront (Table 8). Under a build-operate-transfer type of PPP, a private company designs and finances an infrastructure asset in return for a stream of payments from users and/or the government over the life of a contract, often 20–30 years. In Beijing, customer fees for recycled water are unlikely to be adequate to cover the operating and capital costs of a WRP, so many depend on subsidies from government. For the smaller projects in outlying counties, BWA and the Environmental Protection Bureau are responsible for supervision of the private operators and for paying the treatment fee. While this may be one way to get the asset built initially, it may impose a considerable financial burden on local governments in the longer term and prove to be unsustainable.

Conclusions

The system for water reuse in Beijing is in many respects consistent with the desirable hybrid urban system, combining large facilities, which offer the advantages of economies of scale, ability to cope with variability in quality and quantity of influent water, and an existing

Table 8. Public–Private Partnership Projects.

Project name	District type	Structure	Sewage treatment design capacity (m³/day)	Water reuse design capacity (m³/day)	Developer name (English)	Contract type	Contract length (y)	Operational (year)
Changping Shahe WRP phase 1	Rural		30,000	30,000	Beijing Enterprises Water Group	BOT	29	2013
Changping Baishanzhen WRP	Rural	Bundled	20,000	20,000	Changping Water Affairs Bureau	O&M		2013
Changping Machikou WRP	Rural	Bundled	25,000	25,000	Changping Water Affairs Bureau	O&M		2015
Changping Yangfangzhen north district WRP	Rural	Bundled	8,100	8,100	Changping Water Affairs Bureau	O&M		2015
Changping Shisanlingzhen Huzhuang WWTP	Rural	Bundled			Changping Water Affairs Bureau	O&M		
Changping WRP Phase 2	Rural		30,000	30,000	Beijing Enterprises Water Group	BOT	25	
Changping Xiaotangshanzhen WRP phase 1	Rural	Bundled	70,000	30,000	China Energy Conservation and Environmental Protection Group (CECEPG); YanLong Water (Beijing) Co., Ltd. (YLWC)	BOT		2015
Changping Liucun WRP phase 1	Rural	Bundled	2,000	2,000	CECEPG; YLWC	BOT		2015
Changping Nankouzhen WWTP	Rural	Bundled	20,000		CECEPG; YLWC	BOT		2015
Changping Yanshouzhen WWTP	Rural	Bundled			CECEPG; YLWC	BOT		2015
Changping Shisanlingzhen Changling WWTP	Rural	Bundled			CECEPG; YLWC	BOT		2015
Changping Shisanlingzhen Dingling WWTP	Rural	Bundled			CECEPG; YLWC	BOT		2015
Changping Zhenggezhuang WRP	Rural				Changping Water Affairs Bureau	BOT		
Changping Cuicunzhen WWTP	Rural				Changping Water Affairs Bureau	BOT		
Changping Xingshouzhen WWTP	Rural				Changping Water Affairs Bureau	BOT		
Chaoyang Dongba WWTP phase1	Urban		20,000		Beijing Capital Group DongBa Water Co.	BOT	25	2013
Chaoyang Fatou WRP	Urban		20,000	20,000	Beijing Municipal Road & Building Building Material Group	DBOT	21	2013
Caoyang Beiyuan WWTP phase 1	Urban		40,000	40,000	Beijing Veolia Water Treatment Co.	BOT		2010

Project	Location	Bundled	Capacity	Capacity	Company	No.	Contract	Year
Daxing Tiantanghe WWTP upgrade to WRP	Rural		80,000	80,000	Beijing Jindi Water Co.	29	BOT	2015
Daxing Xihongmen WRP	Rural		40,000	40,000	China Railway 1st Group		BOT	2016
Fangshan Hancunhe WWTP/WRP	Rural	Bundled	4,000	4,000	Beijing Beipai Water Investment (BBWI); Beijing Drainage Group (BDG)		O&M	2013
Fangshan Liulihe WWTP	Rural	Bundled	5,000	5,000	BBWI; BDG		O&M	2013
Fangshan Hebei zhen WWTP	Rural	Bundled	1,200	1,200	BBWI; BDG		BOT	2014
Fangshan Zhoukoudian WWTP	Rural	Bundled	5,200	5,200	BBWI; BDG		BOT	2016
Fangshan Shiduzhen WWTP	Rural	Bundled			BBWI; BDG		BOT	
Fangshan Changyangzhen WWTP	Rural	Bundled			BBWI; BDG		BOT	
Fangshan Qinglonghu WWTP	Rural	Bundled	3,500	3,500	BBWI; BDG		BOT	2015
Fangshan Doudian industry base WWTP	Rural		19,100	19,100	BDG		BT	2014
Fengtai Qinglonghu WRP	Urban		10,000	10,000	Beijing Zhongshui Guofeng Water		BT	2014
Fengtai Lugoujiao WWTP upgrade to WRP	Urban		100,000	100,000	BDG		TOT	2004
Haidian Daoxianghu WRP phase 1	Urban		80,000	80,000	Beijing Enterprises Water Group	30	BOT	2015
Haidian Shangzhuan WRP phase 1	Urban		120,000	120,000	Beijing Origin Water Science and Technology Co. (BOWST)	30	BOT	2015
Mentougou 2nd WRP	Rural		80,000	80,000	BOWST; Beijing Jiu'an Construction Investment (BJCI)	29	BOT	2016
Miyun Xincheng WRP	Rural		65,000	65,000	BOWST	28	BOT	2015
Shijinshan Wulituo WWTP	Urban		20,000	20,000	Beijing Municipal Road & Building Building Material Group	25	DBOT	2013
Shunyi Beiwuzhen WRP	Rural	Bundled	7,500	7,500	BOWST; BJCI	25	BOT	2015
Shunyi Beixiaoying WRP	Rural	Bundled	10,000	10,000	BOWST; BJCI	25	BOT	2015
Shunyi Beishicaozhen WRP	Rural	Bundled	3,000	3,000	BOWST; BJCI	25	BOT	2015
Shunyi Dasungezhuang WRP	Rural	Bundled	7,200	7,200	BOWST; BJCI	25	BOT	2015

(continued)

Table 8. (*Continued*).

Project name	District type	Structure	Sewage treatment design capacity (m³/day)	Water reuse design capacity (m³/day)	Developer name (English)	Contract type	Contract length (y)	Operational (year)
Tongzhou Bishui WWTP upgrade to WRP	Rural		160,000	160,000	Beijing Tongzhou Switch Co.	BOT		2016
Yanqing Zhangshanyinzhen WRP	Rural	Bundled	2,600	2,600	Beijing Capital Group (BCG)	O&M	25	
Yanqing Dayushuzhen WWTP	Rural	Bundled			BCG	BOT	25	
Yanqing Xiangyingxiang WWTP	Rural	Bundled			BCG	BOT	25	
Yanqing Jingzhuangzhen WWTP	Rural	Bundled			BCG	BOT	25	
Yanqing Kangzhuang WWTP	Rural	Bundled			BCG	BOT	25	
Yanqing Badalingzhen WWTP	Rural	Bundled	7,600	7,600	BCG	BOT	25	2015
Yanqing Yongningzhen WWTP	Rural	Bundled	6,000	6,000	BCG	BOT	25	2015
Yanqing Jiuxianzhen WWTP	Rural	Bundled			BCG	BOT	25	
Yanqing County Chengxi WRP	Rural		40,000	40,000	BCG	BOT	28	2015

Note. WRP = water reuse plant. WWTP = wastewater treatment plant. BT = build-transfer. BOT = build-operate-transfer. DBOT = design-build-operate-transfer. O&M = operation & maintenance contract. TOT= transfer-operate-transfer.
Source: Compiled by the authors.

collection network, with small-scale distributed systems in less densely developed areas, which reduce the costs of network installation.

The city is employing a range of technologies, both singly and in combination. In theory, this should allow experimentation and learning at the city level about which technologies are most appropriate to different influent characteristics as well as for different uses, so that the selection of technologies across the hybrid system can be optimized over time.

However, the reuse system in Beijing raises three major concerns: effectiveness, sustainability and value. The first issue relates to the extent to which the treatment infrastructure that has been built and is under construction is being used and is achieving the planned effluent quality. While the average utilization rate of plants in the data-set is fairly high, there is wide variation. It is likely that smaller plants will have lower rates of utilization because of variability in influent quantity and quality, and lower skills and less experience among operators, leading to more system interruptions and incentives to minimize operating costs. Thus the average rate of utilization of plants may decline as the distributed reuse system is extended.

The increase in treatment capacity has preceded investment in distribution networks, so greater attention will need to be paid to this in future plans in order to raise utilization rates. Utilization will also increase if the use of tapwater and groundwater for high-water-consumption / low-value-added activities such as car washing are tightened. The use of WRPs is being held back by fragmented standards. Harmonizing standards and strengthening quality monitoring will raise customers' confidence to switch permanently to recycled water from other sources.

The second major concern is the financial sustainability of the system that has been established. At the moment, the large reuse plants are being financed by the municipal government, while the distributed plants are being financed by district governments. As district governments face stricter financial constraints, they have turned to the private sector to finance investments under build-operate-transfer structures. However, to provide the private investors with adequate returns, the municipal government has had to commit to a stream of subsidies to the private operators to compensate for the suppressed tariffs. Therefore, while PPP may appear to be an effective solution in the short run, it casts into doubt the long-run sustainability of these projects. A risk is that some plants that are built will not be operated. Improving the financial sustainability of water reuse projects should therefore be a priority policy area. This could be achieved by raising the standard tariff for tapwater, as well as the recycled water tariff, to allow WRPs to cover costs while retaining a margin between the two tariffs to incentivize customers to use recycled water.

Finally, the existing hybrid system appears to offer limited value from the supply of reused water: only a small proportion of the water reclaimed is used either for productive purposes or to replace regular piped supplies. While the reasons for this are well known – the absence of networks for distribution of reclaimed water, low tariffs for regular piped supply, concerns about quality, etc. – much greater attention needs to be paid to planning the use as well as the production of reclaimed water. This is particularly urgent given the severity of water scarcity in the municipality, which will only increase with population and economic growth.

Can it be argued that the use of reclaimed water for environmental purposes like river recharge is generating social benefits? Despite the very considerable increase in wastewater collection and treatment capacity in the last decade in Beijing, and the use of reclaimed water for river recharge, it appears that this has not been adequate to compensate for increases in diffuse pollution and in upstream abstraction. Both of these issues will need to be addressed in tandem with the development of water reuse to achieve improvements in water environment quality. Further data collection and analysis would be useful to understand fully the trends in non-point discharges and abstraction in order to assess the potential impact of the use of recycled water for environmental purposes.

Beijing's water reuse system is undergoing a period of rapid development, not just in terms of infrastructure but also in terms of the institutional and regulatory environment. In the overall context of increasing water scarcity in the municipality, Beijing is moving towards an integrated system which will allow the supply of recycled water of different quality levels to be matched to a range of users in multiple locations across the municipal area. Once the planning, pricing and regulatory issues discussed are addressed, Beijing will be able to benefit from the full potential of water reuse.

Acknowledgement

We are grateful to an anonymous reviewer for this point.

Disclosure statement

No potential conflict of interest was reported by the authors.

References

Cosier, M., & Shen, D. (2009). Urban water management in China. *International Journal of Water Resources Development, 25*, 249–268. doi:10.1080/07900620902868679.

Daigger, G. T., & Crawford, G. V. (2007). Enhancing water system security and sustainability by incorporating centralized and decentralized water reclamation and reuse into urban water management systems. *Journal of Environmental Engineering and Management, 17*(1), 1–10.

Daigger, G. T., Rittmann, B. E., Adham, S., & Andreottola, G. (2005). Are membrane bioreactors ready for widespread application? *Environmental Science & Technology, 39*, 399A–406A.

Fane, S. A., Ashbolt, N. J., & White, S. B. (2002). Decentralised urban water reuse: The implications of system scale for cost and pathogen risk. *Water Science & Technology, 46*, 281–288.

Liang, X., van Dijk, P. M. (2010). Financial and economic feasibility of decentralized wastewater reuse systems in Beijing. *Water Science and Technology, 61*, 1965–1973.

Liu, S. & Persson, K. M. (2013). Situations of water reuse in China. *Water Policy, 15*, 705–727.

Shen, D. (2014). Post-1980 water policy in China. *International Journal of Water Resources Development, 30*, 714–727. doi:10.1080/07900627.2014.909310.

Tan, X., Shi, L., Chen, Z. K., Li, T., Ma, Z., Zhang, X., & Chen, R. (2015). Cost analysis of the municipal wastewater treatment plant operation based on 227 samples in China. *Water Supply And Drainage, 41*, 30–34.

Xia, J. (2012). An integrated management approach for water quality and quantity: Case studies in north China. *International Journal of Water Resources Development, 28*, 299–312. doi:10.1080/07900627.2012.668648.

Yang, C., Ren, X. X., & Hu, A. B. (2011). Preliminary study on urban reclaimed water supply model - a case study on Shenzhen Shajing reclaimed water plant. *Proceedings of China Urban Planning Annual Conference, 2011*, 5225–5231.

Yi, L. L., Jiao, W. T., Chen, X. N., & Chen, W. P. (2011). An overview of reclaimed water reuse in China. *Journal of Environmental Sciences, 23*(10): 1585–1593.

Zhang, Y., Chen, X., Zheng, X., Zhao, J., Sun, Y., Zhang, X., … Liao, F. (2007). Review of water reuse practices and development in China. *Water Science & Technology, 55*, 495–502.

Singapore's experience with reclaimed water: NEWater

Hannah Lee and Thai Pin Tan

PUB, The National Water Agency, Singapore

ABSTRACT

NEWater, Singapore's reclaimed water, has enabled Singapore to sustainably meet its growing water demand despite limited land for water catchment and storage. While technology provided this water reuse solution, strong political will, good governance and effective public engagement were key to Singapore's success in supplying NEWater for indirect potable use and direct nonpotable use. A multiple-barrier process including dual-membrane filtration and UV disinfection, complemented by a strict operating philosophy and comprehensive water quality management programme, ensures reliable delivery of good-quality NEWater even as the supply capacity expands.

Introduction

Singapore is a densely populated, highly urbanized city-state with about 5.5 million persons living across approximately 718 km^2 of land, as of 2014, according to the Department of Statistics Singapore (http://www.singstat.gov.sg). Its resultant land scarcity means that although it receives about 2.4 m of rainfall annually, there is limited land available for collection and storage of rainwater. Without alternative sources of supply, Singapore would be highly dependent on imported water from Johor, Malaysia, to meet its water demand of about 1.82 million m^3/day (PUB, 2013).

In 2003, after a period of extensive and in-depth investigation, the Public Utilities Board (PUB) started supplying 'high-grade' reclaimed water, referred to as NEWater, for direct nonpotable use (DNPU) and indirect potable use (IPU). This reclaimed water is of high purity: the quality surpasses standards set out in the World Health Organisation Drinking Water Quality Guidelines (WHO, 2011) and the United States Environmental Protection Agency National Primary Drinking Water Regulations (USEPA, 2009).

NEWater is produced from treated sewage, termed 'used water', that is further purified using advanced membrane technologies and ultraviolet disinfection. This article describes Singapore's approach to water reuse and reviews its experience in developing and growing the supply of NEWater, highlighting the critical success factors behind its implementation and sustained track record of good water quality.

Approach to water reuse

In Singapore, water reuse at the national level is done through two products: a lower grade of reclaimed water known as Industrial Water, and NEWater. Industrial Water was first introduced in 1966 and served as an alternative water source for nonpotable use in industries in Jurong, Tuas, and Jurong Island (refer to Figure 1 for location), to free up potable water for domestic applications.

With the introduction of NEWater in 2003, the approach of using reclaimed water to replace potable water for nonpotable use was continued: NEWater was and continues to be supplied mainly for DNPU in water-intensive industries such as wafer fabrication plants (fabs), power generation and petrochemical industries, as well as in commercial and public buildings for air-con cooling towers.

However, NEWater, which surpasses drinking water standards, is also able to supplement the potable water supply through IPU by injecting NEWater into the reservoirs and allowing it to mix with rainwater before further treatment to potable water. At most times, the NEWater used for IPU constitutes a small proportion of water demand, but this can be increased substantially when larger quantities of NEWater are injected to supplement reservoir supply during dry spells.

NEWater is costlier to produce than Industrial Water because it requires additional treatment steps, comprising reverse osmosis (RO) filtration and ultraviolet (UV) disinfection, to achieve its high quality. The NEWater tariff is SGD1.22/m^3 compared to SGD0.65/m^3 for Industrial Water (PUB, 2014), and both tariffs are set so as to recover the full cost of the respective products. Nevertheless, its wider range of application and ability to enhance water security through IPU make NEWater the preferred option over Industrial Water for water reuse on a national scale. Hence, PUB has expanded NEWater supply capacity over the years to be able to meet up to about 30% of Singapore's total water demand (PUB, 2013).

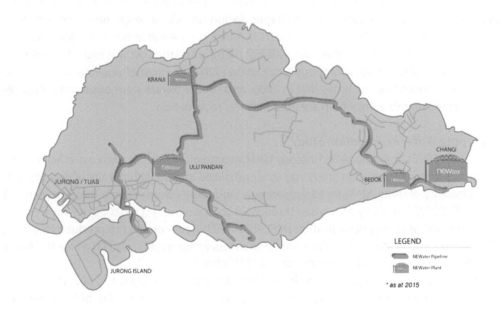

Figure 1. Map of NEWater infrastructure in Singapore.

Importance to Singapore

Since Singapore's independence in 1965, water security has been a permanent consideration for the city-state's leadership in its planning for water resources (Tortajada, Joshi, & Biswas, 2013). In a Water Master Plan developed in 1972, the overall strategy for increasing water supply was to develop surface water schemes while keeping in view unconventional sources, such as water reuse, to be implemented when they became feasible or necessary (Tortajada et al., 2013). With surface water resources almost maximized today, water reuse forms a critical component of Singapore's strategy for water sustainability because it circumvents the limitation of land scarcity for water catchment and storage and reduces susceptibility to variations in rainfall patterns. As a result, Singapore has been able to support greater population and economic growth.

NEWater is also more energy-efficient and cost-efficient to produce than desalinated water due to the lower salt content of the treated 'used water', compared to seawater. While desalination offers great potential to increase supply capacity in the long run, NEWater offers a more cost-effective solution to meet long-term water demand by lowering the quantity of desalinated water required to meet demand. Together, NEWater and desalinated water form important components in Singapore's water supply to provide a sustainable water solution that is resilient to climate change.

Developing and growing NEWater supply

Key milestones

Early exploration of high-grade reclaimed water
In the 1960s and 1970s, Singapore relied heavily on water from local catchments and water that was imported from Johor, Malaysia, under two water agreements. In search of a more sustainable and robust solution to its water needs, in 1974 what was then called the Ministry of the Environment began studying the viability of high-grade reclaimed water with a pilot plant to test various treatment technologies, including RO, ion exchange, electrodialysis and ammonia stripping (Tortajada et al., 2013). Although the water met WHO guidelines for drinking water, technologies then were still not mature, and the cost was high. According to the American Membrane Technology Association (2007), the cost of seawater membranes in 1978 was more than six times that in 2000, and similar trends were observed for brackish-water RO membranes, which are used in water reclamation.

Singapore Water Reclamation Study
By the late 1990s, membrane technology had become more reliable and cost-efficient. The Singapore Water Reclamation Study was conceptualized in 1998 with the aim of determining the suitability of using NEWater to supplement Singapore's water supply through planned IPU (PUB, 2002). The study comprised three main components: (1) a 10,000 m³/day demonstration plant utilizing microfiltration (MF), RO and UV technologies to produce NEWater; (2) a Sampling and Monitoring Programme to assess water quality; and (3) a Health Effects Testing Programme to determine the safety of NEWater.

The demonstration plant was built, and the study ran from 2000 to 2002. Its findings were evaluated by a panel of experts, which concluded that the quality of NEWater surpassed the WHO Drinking Water Quality Guidelines and USEPA National Primary Drinking Water Regulations and it was safe for potable use (PUB, 2002). The panel recommended the IPU approach to provide an environmental buffer as well as allow trace minerals to be reintroduced by blending with reservoir water. Public acceptance was also considered in the recommendation.

Implementation and expansion of NEWater supply

NEWater plants. The first two NEWater plants, in Bedok and Kranji, were completed in January 2003, with a total supply capacity of 72,000 m³/day. A third plant, in Seletar, started operations in 2004 but was decommissioned in 2011 with the closure of Seletar Water Reclamation Plant, which was its source of treated used water. The Bedok and Kranji NEWater plants were expanded in 2009 and 2008, respectively, to more than double their capacity, at 159,000 m³/ day. Meanwhile, in 2007, another plant was commissioned, in Ulu Pandan. This was the first plant developed under a public–private partnership, using the design-build-own-operate model. This allowed a private company (Keppel-Seghers) to supply up to 145,000 m³/ day of NEWater to PUB based on a tendered price. The newest plant was built in Changi in 2010 under a similar design-build-own-operate arrangement, and has a supply capacity of 227,000 m³/day, bringing the total supply capacity to about 532,000 m³/day, or about 30% of Singapore's total water demand. Figure 1 shows the location of NEWater plants in Singapore.

Network infrastructure. NEWater is conveyed to customers and to the reservoirs through a dedicated pipeline network (separate from the potable water system). In the initial years, the supply networks were 'cluster-based'. Pipelines were laid from each NEWater plant to a target cluster of high-demand customers and other nondomestic customers along the pipeline route. For example, Bedok NEWater Factory served mainly Tampines and Pasir Ris wafer fab parks, but also supplied water to commercial and public buildings in the area. NEWater from Bedok NEWater Factory was supplied for IPU by mixing with rainwater in Bedok Reservoir before treatment at Bedok Water Works.

By 2012, an island-wide transmission system was completed, linking the four NEWater plants, five service reservoirs and the various supply clusters. Besides improving supply reliability, the integration of the pipeline network allowed greater flexibility in managing NEWater supply and demand, thus further increasing the resilience of the NEWater system.

Future expansion. With water demand expected to almost double by 2060, Singapore plans to increase NEWater supply to meet up to 50% and 55% of total water demand by 2030 and 2060, respectively (PUB, 2013). Much of this supply will be channelled to industry, which is a major component of the projected demand growth.

Critical success factors

While technology made the production of NEWater viable, strong governmental support and public acceptance made its large-scale implementation successful. Several factors in the planning and development stages were critical in ensuring this.

Strong support from government

Making Singapore self-sufficient in water had always been a top priority for the government. With the formation of the Water Planning Unit in 1971 under the Prime Minister's Office, the resultant 1972 Water Master Plan, outlining plans for local water resource development including water reuse, received the highest levels of political attention (Tan, Lee, & Tan, 2009). This set the stage for strong support to proceed with the NEWater project when the technology was mature enough.

Credible reference projects

While Singapore did not embark on high-grade water reuse after the trial in 1974, PUB and the Ministry of the Environment continued to monitor the tests, studies and projects implemented overseas. By the time NEWater was developed in 2002, water reuse for planned IPU had been practised successfully for more than 20 years in the United States, in areas such as Southern California (Orange County Water District) and Northern Virginia (Occoquan Reservoir, serving the vicinity of Washington, D.C.) (PUB, 2002). In fact, two engineers from PUB and the Ministry of the Environment made a trip to study these projects and others before the decision to develop NEWater was made (Tan et al., 2009).

The positive experiences that other utilities had had with reclaimed water proved critical in winning public confidence. A key message in the public communications surrounding NEWater was that water reclamation was not new and had been successfully practised in other countries, for example the United States, for more than two decades (Tan et al., 2009).

Technology demonstration in the local environment

It was necessary to test the technology in Singapore's environment, to account for differences in climate, the nature of used water, flow patterns, etc. The demonstration plant served as a training ground to build up a team of competent engineers and provided ample opportunity to optimise design considerations and solve operational challenges. For example, chloramination using residual ammonia in the treated used water was found to be an effective means of controlling membrane biofouling with minimal impact on membrane integrity. Also, flow equalization tanks were implemented to manage the diurnal fluctuations in the used water flow, thus optimizing production cost.

Rigorous assessment of water safety

The demonstration plant also provided a continuous supply of water at various stages of treatment, which could be sampled and tested under the Sampling and Monitoring Programme and the Health Effects Testing Programme. Over the two years, some 20,000 test results from seven sampling locations in the plant, covering about 190 physical, chemical and microbiological parameters, were collected (PUB, 2002). This included more than 4500 results for NEWater, which surpassed the WHO and USEPA drinking water standards. The microbiological parameters for NEWater were found to be comparable to or better than PUB drinking water: of the microbiological water quality parameters analyzed, only heterotrophic plate counts were consistently detected in NEWater, and these were lower in concentration than that in PUB drinking water (PUB, 2002). Under the Health Effects Testing Programme, NEWater was tested for short- and long-term toxic and carcinogenic effects in mice and fish, as well as estrogenic effects on fish, and found to have no such effects (Tan et al., 2009).

The comprehensive data on water quality and health effects gave technical experts, political leaders and members of the public confidence in the safety and reliability of NEWater even before supply commenced.

Endorsement by a panel of experts

The formation of a panel of local and foreign experts (including those in engineering, microbiology, toxicology, biomedical science, chemistry, and water technology) to advise PUB and the Ministry of the Environment on the Singapore Water Reclamation Study and review its results lent objectivity and credibility to the findings and recommendations.

Effective public and customer engagement

Despite the technical rigour involved in developing NEWater, Singaporeans had to be convinced that it was safe to use. An extensive public communications plan was rolled out, and public education continues today in schools, at community events and at the NEWater Visitor Centre. The message remains consistent: Potable reuse of water is not a new concept and has been practised successfully around the world; the multiple-barrier treatment process is safe and reliable, and IPU provides a further environmental buffer; NEWater is a sustainable source of water for Singapore.

Engaging the media.   As the media was instrumental in conveying the details of NEWater to the public, their appreciation of the issue was important. In 2002, a study trip was organized for journalists to visit overseas water reuse projects (Tan et al., 2009). The findings of the expert panel were released, and a press conference was held with the experts to help journalists clarify their understanding of the details. Visits to the demonstration plant were also conducted for them to observe the treatment process first-hand.

Community engagement.   Members of Parliament and grassroots leaders were briefed, and they spread the message to the community with the help of exhibitions, posters, brochures and advertisements. NEWater was bottled to allow the public to taste it, and often distributed at community events as well as the National Day Parades.

Change of terminology.   'NEWater' was a carefully chosen product name to emphasize its ultra-clean nature. To further reinforce the idea that NEWater was recycled from used water, which could be used over and over again, similar to that in the natural water cycle, sewage treatment plants were renamed 'water reclamation plants' and sewage or wastewater was termed 'used water'.

NEWater Visitor Centre.   Opened in 2003, the Visitor Centre allowed visitors to view the treatment process at Bedok NEWater Factory from a gallery and understand the science behind it through interactive displays, tours and workshops. It continues to be instrumental in bridging the gap between technical considerations and public perception.

Targeted engagement of wafer fabrication industry.   The wafer fabs were early adopters of NEWater, but this was only possible through close collaboration with PUB. Given its high purity, with low organic and mineral content, NEWater was a good fit for the ultra-pure water (UPW) production process in wafer fabs, but its use was unprecedented. To prove its suitability, PUB, in consultation with the fabs, built a pilot UPW plant to confirm and assure the wafer fabs that UPW produced from NEWater met their specifications.

After they started using NEWater, PUB stayed in close communication with the wafer fab operators, providing daily updates on the water stock level and water quality and addressing their concerns promptly. Over time, the cost benefits of using NEWater became evident. Fewer treatment steps were required to achieve UPW quality when NEWater was used as feed, compared to when PUB potable water was used. According to Tan and Seah (n.d., as cited in Tan, Lee & Tan 2009), the wafer fabs saved about 20% of their chemical cost for UPW production. The successful implementation of NEWater for UPW production also enhanced public perception of NEWater.

Integrated management of the water loop

NEWater closed the water loop, removing the clear separation between water supply and used water treatment. Thus it made sense for PUB to be reconstituted in 2001 to take over the used water and drainage functions from the Ministry of the Environment (Tan et al., 2009). Having a single agency manage all aspects of the water cycle in an integrated manner facilitated the smooth implementation and expansion of the NEWater supply system.

Ensuring the quality of NEWater

A multiple-barrier approach is adopted in design and operation to ensure that the quality of NEWater produced is suitable for DNPU customers and safe for IPU (Seah, Tan, Chong, & Leong, 2008). The approach to operational control and water quality management is based on the observation of trends with respect to the baseline performance of the system. Plant performance and water quality data are reviewed twice yearly by an Internal Audit Panel and audited twice yearly by an External Audit Panel comprising local and overseas experts.

Multiple-barrier approach

Figure 2 describes the multiple barriers involved in the production of NEWater and the role each plays in safeguarding its quality for IPU. Besides the treatment processes, which remove contaminants, source control is critical in ensuring the good quality of used water feedstock for NEWater production. Legislation for proper handling, control and disposal of hazardous substances is in place, and this is enforced through site surveillance. Industry engagement aims to raise awareness of legislation, seek buy-in and share best practices. In addition, on-site monitoring in the sewer network of volatile organic carbons that cannot be effectively removed by the NEWater treatment processes provides early warning to plant operators in the event of any discharge.

Operational control based on real-time data

To ensure that all the NEWater produced is suitable for IPU and customers, there is a need for real-time monitoring of water quality. This is done through the use of surrogate parameters which are indicative of water quality and treatment efficacy and can be monitored continuously online. Plant operators observe the trends of these parameters and take the necessary actions to maintain water quality close to the baseline performance of the plant, which is well within drinking water standards.

The critical surrogate parameters used in the NEWater plant are total organic carbon (TOC) and conductivity. They are measured from the RO permeate and after the product water storage tank. The TOC in NEWater is sensitive to any breach in RO membrane integrity and source water contamination by volatile organic carbons. Conductivity is indicative of the salt content of the permeate water and correlates well with membrane integrity. Using these indicators together, the operator can determine whether deviations in permeate quality are more likely to be due to source water contamination or membrane integrity problems and take the necessary steps to restore water quality. NEWater plants also have the ability to recycle the RO permeate to dilute and maintain product water quality within the baseline range.

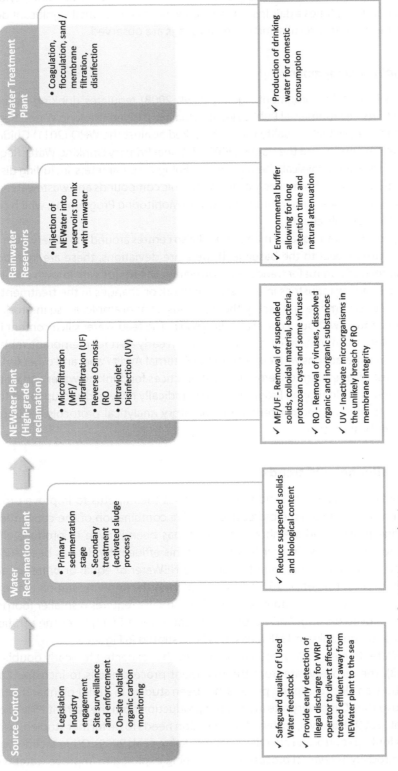

Figure 2. Multiple-barrier approach for NEWater production.

To ensure that online readings are accurate, operators collect samples and perform laboratory analysis three times a day. Instrument servicing is carried out if significant deviations between the laboratory results and online readings are observed.

Water quality management

Singapore's National Environment Agency (NEA, 2008) regulates drinking water quality, according to its Environmental Public Health (Quality of Piped Drinking Water) Regulations 2008. In addition, NEWater quality is benchmarked against the WHO (2011) Guidelines for Drinking Water Quality and the USEPA (2009) National Primary Drinking Water Regulations.

Over 330 microbial, physical, chemical and radiological parameters, including disinfection by-products, pesticides, hormones, persistent organic compounds and wastewater signature compounds, are tracked under the Sampling and Monitoring Programme, which covers the entire NEWater supply chain.

The approach to water quality management also centres around the observation of trends over time, with respect to the baseline. If there are deviations, these are investigated by comparing related parameter trends, and correlating the trends with known events such as the presence of new industries in the sewer network or changes in the treatment process. This is done to determine and rectify the root cause. For example, a rise in product water silica (SiO_2) content without a corresponding rise in the feed-water silica content is usually indicative of RO membrane ageing and serves as an early alert for membrane replacement.

Through its interactions with experts on the External Audit Panel and other forums, PUB keeps abreast of emerging concerns and best practices for drinking water and water reuse. The Sampling and Monitoring Programme is periodically reviewed and updated to include contaminants of emerging concern, and laboratory analytical protocols are improved in keeping with best practices.

Future challenges and trends

As the scale of NEWater production grows, so does the impetus to improve process and cost-efficiency. The membrane bioreactor (MBR), a combination of the conventional activated-sludge process and membrane filtration, has been proven to produce effluent of consistently good quality. Further treatment of this effluent by RO has been successfully pilot-tested in Singapore and found to produce NEWater of equal or better quality than the current process (Qin et al., 2006). PUB is exploring the implementation of the MBR-RO process in Changi Water Reclamation Plant. This process requires a smaller footprint than the current activated-sludge/MF/RO process because the MBR replaces the aeration tanks, final sedimentation tanks and MF tanks. This is illustrated in Figure 3.

For NEWater to meet 55% of total demand, which is expected to nearly double by 2060 (PUB, 2013), the recovery rate from the treatment process needs to increase. Capacitive de-ionization, an electrosorption process, has been studied in a pilot plant and found to be able improve the RO recovery rate in NEWater production from the current 75% to over 90% (Kekre et al., 2009). However, sustainable operation needs to be further studied, particularly to address fouling control and effective cleaning (Kekre et al., 2009). Other technologies, such as the electrodialysis reversal process, are also being explored to improve water recovery.

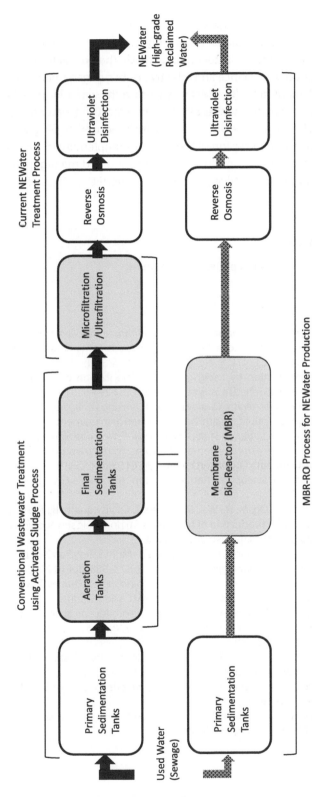

Figure 3. Comparison of current water reclamation process with membrane bioreactor/reverse osmosis (MBR-RO) process for NEWater production.

Conclusion

Over the past 15 years, NEWater has grown from a demonstration-scale project to a proven and well-accepted solution, capable of sustainably meeting more than half of Singapore's water demand in the long run. While technology made the production of NEWater viable, strong political will, good governance and effective public engagement were instrumental to its successful implementation. By effectively closing the water loop, NEWater increases Singapore's resilience to climate change. By constantly recycling water through the system as it is used, the need for large water storage capacity is reduced, freeing up limited land for other uses. The multiple-barrier design of the system, complemented by a strict operating philosophy and comprehensive water quality monitoring and management, has ensured a reliable supply of good-quality NEWater for DNPU and IPU, even as the scale of supply continues to expand.

Disclosure statement

No potential conflict of interest was reported by the authors.

References

American Membrane Technology Association. (2007). *Membrane Desalination Costs*. Retrieved from AMTA Website: http://www.amtaorg.com/wp-content/uploads/6_MembraneDesalinationCosts.pdf

Kekre, K., Tao, G. H., Viswanath, B., Lee, L. Y., Ng, H. Y., Ong, S. L., & Seah, H. (2009). Target of 95% water recovery in NEWater production by using capacitive deionization based process for RO brine treatment. *Proceedings of the Water Environment Federation*, 5294-5301

NEA. (2008). *Environmental Public Health (Quality of Piped Drinking Water) Regulations 2008*. Retrieved from http://www.nea.gov.sg/public-health/potable-water-management

PUB. (2002). *Singapore Water Reclamation Study Expert Panel Review and Findings June 2002*.

PUB. (2013). *Our Water Our Future*. Retrieved from http://www.pub.gov.sg/mpublications/OurWaterOurFuture/Documents/OurWaterOurFuture_2015.pdf

PUB. (2014). *Water Pricing in Singapore*. Retrieved from http://www.pub.gov.sg/general/Pages/WaterTariff.aspx

Qin, J. J., Kekre, K. A., Tao, G. H., Oo, M. H., Wai, M. N., Lee, T. C., Viswanath, B., & Seah, H. (2006). New option of MBR-RO process for production of NEWater from domestic sewage. *Journal of Membrane Science, 272*, 70–77. doi:10.1016/j.memsci.2005.07.023.

Seah, H., Tan, T. P., Chong, M. L., & Leong, J. (2008). NEWater – Multi Safety Barrier Approach for Indirect Potable Use. *Water Science & Technology: Water Supply, 8*, 573–588. doi:10.2166/ws.2008.130.

Tan, Y. S., Lee, T. J., & Tan, K. (2009). *Clean, Green and Blue: Singapore's Journey Towards Environmental and Water Sustainability*. Singapore: Institute of Southeast Asian Studies.

Tan, P., & Seah, H. (n.d.). Impact of NEWater as feedwater for the production of ultra-high-purity deionised water and manufacturing process. *Future Fab International Issue 16 – Process, Gases, Chemicals, and Materials*. (as cited in Tan et al, 2009)

Tortajada, C., Joshi, Y., & Biswas, A. K. (2013). *The Singapore Water Story: Sustainable Development in an Urban City-State*. Abingdon: Routledge.

USEPA. (2009). *National Primary Drinking Water Regulations*. Washington, DC: United States Environmental Protection Agency.

WHO. (2011). *Guidelines for Drinking Water Quality (4th Ed.)*. Retrieved from http://www.who.int/water_sanitation_health/publications/dwq_guidelines/en/

Overcoming global water reuse barriers: the Windhoek experience

P. van Rensburg

Department of Infrastructure, Water and Technical Services, City of Windhoek, Namibia

ABSTRACT
Water scarcity is a reality, with a recent UN report estimating that about half of the global population could be facing water shortages by 2030. This has focused attention on existing sources and what could be done to maximize potential. Water reuse, in particular direct potable reuse (DPR), has enjoyed a somewhat turbulent history globally. Despite this, the City of Windhoek has been practising DPR for more than 45 years, and this commentary presents globally accepted barriers standing in the way of DPR and attempts to explore ways to overcome these given the experience in Windhoek.

The value of water

One thing the value of water is not is the amount you pay to water utilities to have it piped and delivered to your home. This would constitute the *price* of water, which is more often than not determined by the environment and region you reside in. The rate or levy you pay for water does not necessarily reflect the *value* of this precious resource.

In commercial terms the value of water is often linked to its primary use. For instance, certain activities, such as mining or agriculture, can economically link its worth to the value of the raw material extracted or the crops produced (Bolton, 2014). Similarly, tourism activities can be coupled to the income generated from the availability of water to sustain these activities. Or rather, these consumers can, provided they are savvy enough, measure the value of water against the potential financial losses to be suffered from an insufficient supply of water and the often hidden risk to the sustainability of their businesses. On the other hand, a typical urban consumer who pays a per-unit rate for water seldom thinks about its significance and simply expects the water to be there whenever they turn on a tap. This type of consumer is unlikely to factor the economic value of water into their daily budgetary allowance; i.e. they would generally not weigh the cost of having a luxurious bath against the cost of another tangible item that they could have purchased (Paul, 2012).

In any case, the full value of water cannot be measured without considering the broader environmental and social factors that often recede into the background when weighed against the more prominent, and often promoted as more significant, economic influences. The true value of water is defined by not having any water for even the most basic of needs. This is the value realized when the gap between supply and demand becomes the dominant market factor.

Facing water scarcity

A recent report (WWAP, 2015) estimates that about half of the global population could be facing water shortages by 2030, when demand is predicted to exceed water supply by approximately 40%. In fact, water scarcity is considered one of the primary challenges facing many countries, and the world, in the twenty-first century; and given current climate change forecasts it could be even worse than currently predicted.

> Water scarcity already affects every continent. Around 1.2 billion people, or almost one-fifth of the world's population, live in areas of physical scarcity, and 500 million people are approaching this situation. Another 1.6 billion people, or almost one quarter of the world's population, face economic water shortage (where countries lack the necessary infrastructure to take water from rivers and aquifers) (UNDP, 2006).

Though there is theoretically at present still enough freshwater on the earth to sustain 7 billion people, it is unevenly distributed, with a high degree of waste and pollution. Of particular concern is the fact that water usage has reportedly been growing at more than twice the rate of population increase. Add the effects of climate change to the already daunting scenario and the world is in dire need of innovative solutions to address the problem.

Water reuse: the hesitant solution

Against this background, water reuse has, in some areas, been established as an effective mechanism to combat water scarcity. The current global situation is likely to increase the level of interest in reuse as a solution for both mitigating the environmental impact of wastewater disposal and creating an additional supply source. However, even though experts agree on the significant potential for the further development of water reuse projects, the reality on the ground paints an entirely different picture.

Amidst a lack of truly reflective global data it was reported that in 2013 countries in North America treated 61 km^3 of wastewater (1 km^3 is a billion m^3), with a mere 3.8% of the treated wastewater being reused. European figures are not much different, with a 2006 report indicating a volume of 964 million m^3/y of treated wastewater being reused, which amounts to only 2.4% of the treated urban wastewater effluents (UNU-INWEH, 2013; European Commission - Environment, 2013; UNESCAP, 2012).

In contrast to the poor global reuse statistics with predominant use in the agriculture sector, Windhoek, the capital of Namibia, has been successfully treating wastewater for direct potable use for more than 45 years. Whilst the reuse of water may not be an appropriate solution in all circumstances for a variety of reasons, surely the question should be asked as to why, despite a sharp increase in water-scarce regions around the globe, there appears to be a reluctance to implement much-needed water reuse schemes.

Perceived barriers to water reuse

It is clear that despite a number of water reuse applications already developed and established in many countries, there are still a number of barriers which prevent the widespread implementation of water reuse on a truly global scale. On the positive side, the looming global water crisis has seen a definite increase in the level of interest, especially in the less conventional practice of direct potable reuse (DPR). Recent times have seen much dialogue on a variety of global platforms being expended on identifying the barriers to water reuse and exploring ways and means to overcome these (Sanz & Gawlik, 2014). Reviewing the published findings on the obstacles hindering global water reuse, the following can be identified as the primary contributors, in particular insofar as reuse for potable means is concerned:

(1) Public perception/acceptance
(2) Appropriate/standardized technical solutions
(3) Monitoring/management of health considerations and risks
(4) Reuse not being a part of integrated water supply strategies
(5) Water pricing and business models
(6) Regulatory and policy issues (lack of local/regional/global standards and best practices).

The implementation of water reclamation and reuse therefore does not only suffer from technical barriers but also faces other, often far more intimidating challenges, such as limited institutional capacity, lack of financial incentives, and public perceptions of water recycling, reclamation and reuse. In fact, evaluating the above, it can be seen that essentially all six proposed barriers can be categorized into one of only two groupings – technical and nontechnical – with the latter representing the bulk of the issues identified. Allocating each identified barrier type to its designated category or grouping yields Table 1. Knowing the enemy, however, presents only the first step in the process. These barriers will have to be overcome if wastewater reuse is to be accepted and implemented on a larger, more effective global scale than at present.

While these objections certainly raise some very worthy issues to be addressed, the reality is that many reuse projects have been successfully practised for decades whilst confronting these challenges on a daily basis. Granted, a number of the objections are directly linked to the possible impacts in the case of failure, and in this regard the stakes in all reuse projects are most definitely not equal. For instance, the associated health risks in a landscape irrigation project are surely not on a par with those in a DPR scheme.

Secondary to this is the fact that confidence and trust are built slowly over time. People by nature are wary of the unknown, and novel ideas are often treated with contempt until proven trustworthy. It could therefore perhaps add value in the quest for wider global reuse practice to match the existing barriers against the standard practice and mitigating efforts employed in an established reuse application with both high risk factors and an established, long-term success record.

Table 1. Technical and nontechnical barriers to expanded global water reuse.

Technical barriers	Nontechnical barriers
Appropriate technical solutions Managing health risks	Public perception/acceptance Incorporation of reuse into integrated water supply strategies Economic issues Regulatory/policy issues and institutional capacity

Perceived barriers versus the reality

We now return to the question as to why, 45 years after the technology was successfully implemented, and despite a huge database of information charting performance over this period, Windhoek remains one of only a handful of cities in the world practising direct potable reclamation. Yes, there are a number of known barriers, and it may very well be that elsewhere the pressure has never been high enough, with other viable alternatives always reigning supreme in comparison. However, surely the time has come when alternative options are no longer a given and the survival of humans in many places could very well depend on the provision of potable water from wastewater reclamation. It may therefore be prudent to carefully consider the objections to wastewater reuse, one by one, set against a shining example of reuse in its purest form, that of direct potable reclamation in Namibia.

DPR in Windhoek, Namibia, as a reference project

Windhoek (population approx. 370,000 – Namibia Statistics Agency (2013)) is the capital city of Namibia, the most arid country in Sub-Saharan Africa. It is therefore to be expected that the city is no stranger to water scarcity, and as such has run out of conventional supply sources (groundwater and surface water reservoirs) on a number of occasions, with the early 1960s as a definite turning point (Brand, 1962). It was during that time that the city inaugurated a new wastewater treatment plant capable of producing high-quality effluent. Over the period of 1964 to 1968 the Windhoek City Council, in collaboration with the National Institute for Water Research and the Council for Scientific and Industrial Research in South Africa (at that time Namibia was administrated by the Republic of South Africa under a UN mandate), carried out specific research, including a pilot study, aimed at the direct potable reclamation of treated wastewater (Van der Walt, 2003). In 1968, during a time of severe water shortages, this became a reality when the first reclamation facility started operations, producing high-quality effluent distributed for direct potable consumption.

Ever since, the scheme has been successfully operated, encompassing a number of changes over the years to the original design, effected as dictated by ongoing research and technology improvements. During the middle-to-late 1990s another severe drought prompted the design of a new reclamation facility. Again, a period of stringent pilot tests in conjunction with actual production data dictated the final design, which was subsequently constructed and commissioned in September 2002 (Du Pisani, 2004). The initial reclamation plant is still employed as a water reclamation facility for lower-quality effluent, providing irrigation water for numerous public parks and sport fields in the city.

Reliable technological solutions (technical barrier)

Water reuse has been around in one form or another for a very long time, with planned reuse schemes for treated effluent in existence for the best part of a century. As a result, the technology required for reuse is fairly well established, yet surprisingly many areas in the world lack the basic infrastructure necessary for the collection and treatment of wastewater, which directly affects the possibility of implementing water reuse schemes. In determining the technology required, the quality and quantity of wastewater need to be balanced against the intended purposes of reuse. What is considered appropriate technology is therefore primarily determined by the required standards for a specific reuse application. Over time by far the most prevalent application of water reuse has been linked to the irrigation of agricultural and/or urban environments, with both reuse standards and applicable technology fairly well established. This is not so much the case in water reclamation for potable reuse, even though it has been actively employed in the augmentation of potable water supply sources since the late 1960s. However, in spite of a variety of reuse schemes practised successfully around the world for decades, yielding a substantial amount of high-quality and reliable data, the perception persists that proven, robust and cost-effective technological solutions are lacking.

Looking closer at the rapid development of available technology over the last decade in particular, several breakthroughs in wastewater treatment and reclamation for water reuse happened as a result. The most prominent is certainly the refinement of membrane technology, which has emerged as a significant innovation for treatment and reclamation, as well as a leading process in the upgrade and expansion of wastewater treatment plants. An increase in quality and operability, coupled with a sharp decrease in cost, led to a rapid rise in the volume of wastewater globally that is treated with membrane technology to exceptionally high standards ideally suitable for reuse purposes. Today several categories of membrane technology are in existence, typically removing, with great reliability and consistency, contaminants other processes cannot (or struggle to) remove. These include dissolved species; organic compounds; nutrients; colloidal and suspended solids; and human pathogens, including bacteria, protozoan cysts, and viruses (Daigger, Lozier, & Crawford, 2006).

Membrane technology has unquestionably brought about huge improvements in the quality of treated and reclaimed water. On the downside, this technology increases operational and maintenance requirements, in particular the cost of operations and the required skill level of operating personnel (Khalil, N/A).

While on the subject, it is important to note that the maintenance and operation of treatment facilities for water earmarked for reuse require a certain level of accuracy and precision to ensure consistent quality of effluent that meets the required reuse standard and hence is suitable for this purpose without affecting the risk profile of the relevant reuse scheme. Thus, careful attention has to be given to reliability features during the design, construction and operation of these facilities. Apart from the technical design, the maintenance program and the skill level of operating personnel are seen to play important parts in achieving overall reliability of the installation. It therefore goes without saying that a lack of skilled operating personnel and general maintenance shortcomings, such as shortage in supply of general maintenance items and consumables, can act prohibitively against the establishment of water reuse schemes.

Reviewing the above in the context of Windhoek and the Goreangab Water Reclamation Plant (GWRP), the technical design was based on a treatment objective established in terms of reviewing a number of water quality guidelines in existence at the time. A review of the above, combined with a number of other guideline restrictions for individual parameters, led to a combined Final Water Quality Design Guideline that was set as the primary treatment objective (FMG Goreangab Joint Venture, 1998a).

For contaminants which are easily measured and for which reliable health data are available, a maximum contaminant level was specified. Alternatively, as with most microbiological contaminants, a specification of the treatment technique was incorporated in the form of *Ct* values (concentration × time). Rather than measuring all possible microbiological contaminants and attempting to keep each under a certain defined limit, the plant was carefully designed to ensure a specified *t* (time) and plant operation is geared towards maintaining a certain *C* (disinfectant concentration) to give the desired *Ct* product.

Taking into consideration the raw water origin and quality in relation to the overall treatment objective, a technical design based on the so called multiple-barrier concept and developed through extensive pilot- and bench-scale research was devised. The term 'multiple barriers' is often used in relation to water reuse, although its exact definition and/or meaning can vary. In the case of Goreangab, this was developed to literally mean more than one barrier or treatment process per defined pollution parameter and/or grouping (Van der Walt, 2003). Given the risk profile of each defined grouping, a number of barriers were proposed to ensure a technical design that would prove sufficient to remove these even under trying circumstances. A holistic approach was adopted in which not only conventional treatment barriers were concerned. Subsequently, barriers were distinguished and classed as follows:

Nontreatment barriers

- Separation of industrial and domestic wastewater, with the latter exclusively used for the purpose of potable reuse – this includes the strict policing of any industrial-type effluents within the catchment for domestic wastewater
- Continuous and rigorous quality monitoring regime (on-line and/or laboratory) for both raw and final treated water
- Blending of reclaimed water with water from conventional sources to a max. of 35% before distribution to consumers – the reclaimed final water target for dissolved organic carbon (DOC) is 3 mg/L, based on the guideline for total organic carbon from Water Factory 21, in California, where the injected water contains less than 1 mg/L (Williams, 1996)

Treatment barriers

- Permanent presence as part of process train, although not necessarily a physical barrier or 'dead stop'
- Regarded as either partial or complete

Operational barriers

- Backup/additional capacity for existing processes, e.g. addition of powdered activated carbon

Table 2 illustrates the above concept in practice, indicating contaminant/pollutant grouping, number of barriers and associated treatment processes (Menge, 2006).

Considering all of the aforementioned factors, the proposed concept was realized in the form of an intricate process train. On-site treatment processes can be briefly explained as follows (the complete treatment process train is diagrammed in Figure 1):

(1) Powdered activated carbon dosing as a standby process / optional barrier in the event of failure of key processes such as ozonation, granulated active carbon filtration and membrane separation
(2) Pre-ozonation for the oxidation of iron and manganese

Table 2. Process selection aimed at matching treatment objectives in drinking water reclamation.

Treatment objective	Required barriers		Required/proposed process steps(P = partial, C = complete)
	Partial	Complete	
Aesthetic		2	1.C> CD + DAF + SF 2.C> UF
Microbiology and virus	1	3	1.P> GAC 1.P> ASP-BNR-MP for TC, TN and TP 2.P> CD + DAF + SF 1.C> O_3 2.C> UF 3.C> breakpoint chlorination
Protozoa: *Giardia* + *Cryptosporidium*		3	1.P> MP 1.C> CD + DAF + SF 2.C> O_3 3.C> UF
Organics	2		1.P> CD + DAF + SF 2.P> O_3 + BAC + GAC 3.P> PAC
DBPs	2		• Enhanced coagulation • Delay chlorination in process train & reduce dosage
Residuals: Fe, Mn	2		1.P> NaOH + MnO_4 + SF 2.P> O_3 + BAC + GAC
Stability		1	1.C> NaOH
Nitrogenous and organic constituents	1		1.P> ASP-BNR-MP for TC, TN and TP

Notes: **CD + DAF + SF** = chemical dosing + dissolved air flotation + rapid sand filtration, including slow start + filter-to-waste (removal of suspended solids and partial removal of dissolved solids). **O3 = ozonation (oxidation/disinfection). GAC** = granular activated carbon (adsorption of micropollutants). **PAC** = powdered activated carbon (adsorption of micropollutants). **O3 + BAC + GAC** = ozonation + biological activated carbon + granular activated carbon (oxidation + biological degradation + adsorption of micropollutants). **UF** = ultrafiltration (removal of particles). **NaOH** = sodium hydroxide (stabilization of corrosive water). **NaOH + MnO4 + SF** = sodium hydroxide + permanganate (oxidation of iron and manganese + subsequent precipitation). **ASP-BNR for TC, TN and TP** = biological nutrient removal activated sludge process for total carbon, total nitrogen and total phosphorus removal). **MP** = maturation ponds with > 5 day retention. **DBPs** = disinfectant byproducts.
Source: Menge (2006), reproduced with permission.

(3) Acid dosing as part of ferric chloride dosing for when pH needs adjustment to facilitate enhanced coagulation for better organic precipitation

(4) Ferric chloride and polymer (optional) dosing system for flocculation

(5) Two flocculation units

(6) Dissolved air flotation

(7) Rapid sand filtration with upstream dosing of potassium permanganate and caustic soda to facilitate manganese oxidation and removal across the filters and stabilization

(8) Ozone disinfection as a barrier against viruses and bacteria as well as the oxidation of organics making them more susceptible to removal in the downstream activated carbon processes

(9) Three-step granular activated carbon treatment facility – first biological carbon facility (protected through hydrogen peroxide dosing as pretreatment) followed by two granulated active carbon units for adsorption purposes

(10) An ultrafiltration unit process as a physical barrier

(11) Chlorine dosing, guaranteeing a residual of at least 1 mg/L free chlorine for final water dispatched from the plant

(12) Final stabilization (caustic soda) to obtain a positive calcium carbonate precipitation potential of 4 mg/L.

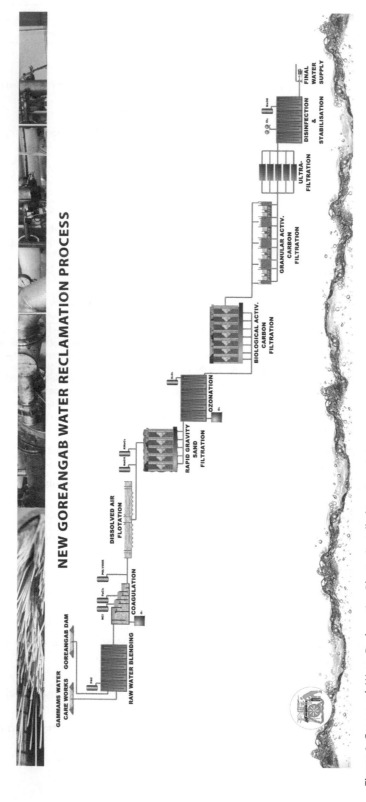

Figure 1. Goreangab Water Reclamation Plant – installed treatment steps.

The GWRP is operated by a crew of skilled personnel, all thoroughly trained local residents, on a three-shift basis. The plant is fully automated, with a modern instrumentation and control system that ensures safe automatic operation with limited manual intervention. Nonetheless, in the event of faults arising, all plant components are automatically transferred to an operating condition which reliably excludes any risks to people and the environment. In this regard all safety and protection functions for plant and equipment will operate fully automatically and reliably. This also applies to the engagement and switch-over functions for redundancies and reserve equipment. For the purpose of operation, a single operator is on duty at all times, with only two assistant operators to attend to fault reports generated by the fully integrated supervisory control and data acquisition (SCADA) system. The control system also generates full historic records in the form of event logs, sequential fault registers, important trends (in graphical representations), alarm time and status, operator intervention, etc.

The entire installation is supported by a competent team of qualified technical staff, ensuring continuous and timely scheduled maintenance as well as fast reactive maintenance in the event of mechanical or treatment failure. In addition, an effective administration and procurement system guarantees a reliable supply of spare components and consumables to sustain a robust, reliable and efficient water reuse facility.

Managing health risks (technical barrier)

Perhaps linked to public acceptance in a major way, the effective management of health risks associated with water reuse is often slated as one of the biggest barriers. Effective management thereof would include identifying all potential hazards and hazardous events of the water reuse scheme and assessing the level of risk they pose to human and environmental health. Of equal importance is uncompromised safety, proven in terms of both short- and long-term effects through extensive sampling and monitoring coupled with appropriate toxicological studies.

In Windhoek, monitoring forms an integral part of the overall potable water supply, but in particular pertaining to the reclamation scheme, to ensure absolute safety of the drinking water and to maintain public trust as part of overall public acceptance as alluded to in the paragraph above. Such monitoring covers the full water cycle. At the GWRP, continuous monitoring is performed through on-line instrumentation as well as the collection of on-line 24-hour composite samples. Instant feedback is obtained through on-line instrumentation placed downstream of all major treatment processes, while composite samples are analyzed at the city of Windhoek's Gammams Laboratory. For this purpose the city has invested heavily over the years in establishing state-of-the-art facilities, including modern analytical equipment and skilful and well-trained personnel. Routine tests conducted at the laboratory include physical, inorganic and organic chemistry, as well as microbiology and viral indicators.

As part of the health risk assessment and monitoring program, advanced tests are conducted by external laboratories in the region (South Africa) and in Europe. These include virology, parasites, toxicity, mutagenicity, natural organic matter, pharmaceutical substances, endocrine disrupting chemicals, disinfectant byproducts, pesticides, algae toxins, and taste and odour compounds.

In light of the ever-increasing analytical capacity allowing for the detection of known contaminants at much lower levels, and the discovery of 'new' contaminants, the water reclamation facility in Windhoek has thus far withstood intense scrutiny through numerous local (internal) and international studies. Nonetheless, the authorities remain acutely aware of the origins of the raw water; hence additional parameters such as total organic carbon and in particular DOC are employed as surrogate indicators over and above derived final water quality standards.

In safeguarding both technical barriers examined above, the GWRP is fully ISO 9,001:2008 certified to ensure that quality control in all areas of operation is optimally managed and maintained.

Although much can be learned from the above practices, health concerns are arguably set to remain the biggest barrier to water reuse, in particular for the purpose of human consumption and consequently, gaining much-needed public acceptance of proposed future reuse schemes.

Public acceptance (nontechnical barrier)

While adequate technology and water quality monitoring systems can promise the delivery of safe, sufficient and secure water for reuse practice, even to the extent of providing drinking water through DPR, gaining public acceptance appears to be a major hurdle (Cain, 2011). It is therefore the author's firm opinion that the technical barriers, often described as the 'core' issues, must not be seen in isolation in evaluating the viability of reuse schemes, in particular DPR. It should further be noted that public perception is almost certainly linked to education and an intrinsic knowledge about the proposed activities available for scrutiny by the members of the general public.

Ironic in the context of the intense public scrutiny of DPR is the fact that unplanned indirect potable reuse, whereby treated municipal wastewater, and sometimes untreated or poorly treated agricultural or industrial wastes, are returned to a water body upstream of an off-take for a conventional drinking water treatment plant, is being practised to this very day in many places in the world. Yet this practice is considered acceptable despite the ever-deteriorating quality of source water bodies from which potable water is abstracted.

Planned indirect potable reuse, whereby an environmental barrier in the form of groundwater or surface water is incorporated, is an established form of water reuse, with NEWater in Singapore (Lee & Tan, 2016) and Water Factory 21 in Orange County, California, perhaps the best-known examples. The reason planned indirect potable reuse is not considered to pose a health risk is that the treated wastewater benefits from natural treatment during storage in surface water and aquifers and is diluted with 'ordinary' river/ground water before abstraction to ensure good drinking water quality (part of a multi-barrier approach in the water safety plan). The storage time provides a valuable buffer to measure and control quality. DPR, however, is almost a closed-loop system, with limited storage and a shorter buffer time, therefore increasing the risk.

In the above context the absence of any natural reservoir or perennial rivers in the region severely reduced the options for potable reuse in the case of Windhoek. This, coupled with severe water shortages and a lack of viable alternative water supply sources, saw DPR considered at an early stage. These were powerful initial drivers for the establishment of the scheme, even though it is highly doubtful that the initial facility would be acceptable anywhere in the world today. Over the years, the public have become so used to having reclaimed water as part of their water supply that many have either forgotten or simply are not aware of this fact.

A study conducted in 2011 entitled *Public Perception of Windhoek's Drinking Water and Its Sustainable Future* presented some rather interesting findings (Boucher, Jackson, Mendoza, & Snyder, 2011). Among others, it was found that 93% of the residents surveyed reported drinking tap water, and about 84% reported the water to be of acceptable drinking quality. However, as to the source of their drinking water, only 44%, less than half, confirmed knowledge of the existence of the GWRP. This is surprising, as approximately 90% of households surveyed considered the water to be absolutely safe for human consumption despite many not knowing the exact origins of their tap water. The survey also indicated that residents who were either born in the city or had lived there for longer than 10 years were much more aware of the reuse component in their drinking water, leading to renewed efforts by the city to disseminate this knowledge in an effort to educate the general population. In this regard youngsters and schoolchildren are specifically targeted as part of science projects, including guided tours of the facility.

Despite the Windhoek experience, public acceptance remains an important cornerstone for the establishment of a water reuse project anywhere in the world, despite vast technological advances. Pursuant to this, authorities in Windhoek are acutely aware that a single public health incident can destroy in the blink of an eye public trust gained over decades.

Reuse considered as 'conventional' water supply source (nontechnical barrier)

It is abundantly clear that tackling the challenges posed by the imminent global water crisis will require an innovative approach to the problem in order to define solutions that are both technically viable and sustainable in the long run. In this regard it is absolutely crucial that wastewater be seen as the valuable resource it is and incorporated by planners and other technical personnel in their assessment of the overall water cycle and available supply sources. This is already commonplace in traditionally dry and water-stressed areas around the globe, where advanced water reuse has become an inseparable part of total water management. Countries and regions falling on the other end of the spectrum, with traditional abundance of freshwater, will soon be forced by increased population pressure, economic demand and pollution of conventional sources into rethinking their approach to sustainable water cycle management. Here new thinking in water resource management is needed, whereby planners and engineers recognize wastewater as a supply source to be harvested as a means of augmenting supply, provided that adequate treatment is applied. This approach is however still severely hampered as a result of insufficiently integrated water management, largely due to the fragmentation of responsibilities into separate water and wastewater authorities in many parts of the world. This leads to a lack of communication and cooperation between the relevant stakeholders, often resulting in an inability to manage the whole water cycle as a continuous chain.

Located in an arid environment, the city of Windhoek adopted this approach many years ago; hence the establishment of the DPR scheme in the late 1960s. Currently the local authority also operates an irrigation-water supply scheme consisting of suitably treated wastewater for the irrigation of public parks and sport fields. In fact water reuse has become such an integral part of the overall water supply scheme that very few planners, engineers or developers of water or related infrastructure do not incorporate this aspect into their project development to some extent. In addition, a portion of the reclaimed water, subject to availability and blended with surface water to strictly prescribed proportions, is used to artificially recharge the existing aquifer as part of the Windhoek Managed Aquifer Recharge Scheme. In this regard it is of importance to note that the Local Authority in Windhoek employs a single department for both the management of potable water supply and wastewater collection and treatment, which greatly assists any form of integration and collaboration across these two sectors.

In order to implement integrated and sustainable water management, it is necessary to bridge the tight but artificial compartments of water supply and sanitation. Too often, water reuse is excluded from possible integrated water management scenarios due to the misperceptions of stakeholders. It is therefore clear that an emphasis on the education of responsible individuals and perhaps a rethink in the structuring of institutional authorities is required in order to stimulate an integrated approach to water resource management, especially in newly water-stressed regions around the globe.

Economic considerations (nontechnical barrier)

Typically, to encourage its use, reclaimed water for nonpotable use is priced just below the cost of potable water, but this can seriously affect cost recovery, and hence the overall feasibility of potential reuse projects. Similarly, potable water from conventional freshwater resources generally lacks full cost recovery and is not representative of a number of external costs associated with its abstraction, purification and conveyance (in the event of nonlocal supply sources). This affects the water market, to the effect that it creates an artificial price difference between reclaimed water and freshwater, often leading to the former's being disregarded as a feasible alternative for economic reasons.

While in some instances reclaimed water systems are developed in response to effluent disposal needs and customers are encouraged to make use of an 'unlimited' supply at little to no charge, more often than not economic feasibility plays a major part in the establishment of such a system. Calculating the true cost of a water reuse system typically includes a significant capital cost portion (improvements to treatment facility, conveyance infrastructure, storage, etc.) in addition to operating cost, which involves determining those treatment, distribution and monitoring components that are directly attributable to the system. Recovering these costs from the consumer might prove challenging, especially in the instance where potable water is readily available and insufficiently priced. As a result, it is common for the cost of reclaimed water to be based on a percentage of the cost of potable water, regardless of the true cost of providing the service. But with DPR the cost of reclaimed water can be recovered at the cost of potable water regardless of the input cost. As mentioned earlier, a major contributing factor to the economics of water reuse can be regions where the water and wastewater utilities are split. Here the potential drop in revenues associated with a reduction in potable water use as a result of the implementation of

Table 3. Windhoek water tariffs.

Potable water (bulk supply)	Conventional surface water	NAD 15.45/m³
Irrigation water (customer supply)	Reclaimed water	NAD 11.19/m³
	Comparable potable supply	NAD 90.00/m³
	Reuse (commercial)	NAD 10.04/m³
	Reuse (public)	NAD 3.81/m³

a reuse system leads to a loss of revenue, in addition to the cost of the reuse scheme, that can be particularly challenging for these utilities.

In the Windhoek scenario, where all conventional supply sources of potable water within a 500 km radius have been exhausted, the value of reclaimed water is directly linked to the cost of providing an alternative. Current conceivable alternatives include the conveyance of water from the Okavango River, over 800 km away, or desalinated seawater over an equally challenging distance and to an altitude of 1700 m AMSL. By comparison, the cost of developing these schemes and the immediate need for additional sustainable supply sources by far eclipsed the cost of reclaiming wastewater as alternative. It is therefore not a question of what the cost of reclamation is, but rather how much water can be reused. Nonetheless, reclaimed water at the GWRP is currently being produced at a cost about 37% lower than the cost of potable water from surface water sources. Having said that, the cost of potable water from the surface water reservoirs is not considered cost reflective, as little investment in the way of developing new water supply sources has taken place in the last two decades.

In addition, semi-purified water for the irrigation of public parks and recreational facilities is being sold at a considerably lower price compared to potable water. Here a distinction is made between consumers purchasing the water for commercial activities and public recreational consumers, with the former paying more than twice as much. Even so, the cost of semi-purified water at commercial tariff is about 10% of the potable water tariff, based on an average daily consumption of 5 m^3 per day. For the irrigation of public recreational facilities, the cost for the use of semi-purified water compared to a potable alternative is almost negligible, comprising about 4% based on the scenario illustrated above. It is however important to note that the tariff for semi-purified water is also currently below cost recovery, with current investment expected to lessen the margin (Table 3).

Water economics, in particular in a water reuse scenario, is complex and very dependent on local circumstances in terms of the supply and demand and the availability of a sustainable conventional supply source. Nonetheless, it is clear that with increasing water scarcity the dynamics are set to change, with the cost considerations in terms of reuse expected to become much more favourable in the face of growing population pressure, diminishing supply and greater pollution threats.

Policies and regulation of water reuse (nontechnical barrier)

Clear guidelines on available technology, universally acceptable policies on health and environmental safety and risks bring about a lack of confidence amongst authorities to make well-defined and substantiated decisions on the future of reuse projects. Limited institutional capacity to formulate and institutionalize enabling legislation and to subsequently conduct adequate enforcement and monitoring of water reuse activities reinforce this as a formidable barrier to overcome.

Most countries where substantial water reuse has been practised for an extended period operate within some sort of a regulatory framework (whether formal or not) addressing the core issues and practicalities of planning, implementing and even in some cases the operation associated with water reuse schemes. In this regard the US Environmental Protection Agency in 2004 published extensive guidelines covering most topics on the theme of water reuse, though they do not constitute enforceable legislation (USEPA, 2004).

Similarly, in Namibia, guidelines for water reuse have been published by the authorities, with mention of water reuse in some applicable legislation, but no specific reuse legislation is in existence (Department of Water Affairs & Forestry [DWAF], 2012). On a local authority level, the city of Windhoek, having practised reuse extensively for nearly half a century, has established internal policies and guidelines governing the reuse of water for internal applications, but these lack inclusion in a formal regulatory framework for national implementation and/or international reference.

Globally, particularly in water-stressed regions, regulations are needed to oblige authorities to assess the contribution water reuse can make to the management of the overall water cycle in accordance with a clear framework for managing health and environmental risks. In addition, a global 'best practices' policy would contribute a great deal to installing confidence in authorities in areas where reuse is not considered as a result of the unknown consequences of decision making in an unregulated environment. On the other hand, it should be realized that overly strict standards and regulations could easily limit water reuse to a few instances of legal reuse and a high number of illegal – and thus unmonitored – reuse practices.

Future considerations and concluding remarks

There is no doubt that water reuse is a topic of significant relevance at present, with its importance expected to grow exponentially in the near future given current predictions on the global freshwater supply situation. Worldwide, the water reuse sector is expected to migrate from agricultural irrigation towards higher-value exploitations, mostly in municipal applications, such as drinking water supply, industry, and landscape irrigation reuse. It is further estimated that water reuse will grow more quickly than desalination in percentage terms and may even exceed the latter in volumetric terms in the not so distant future. In particular, DPR is expected to gain significance in the midst of ever-increasing demand driven by population growth and associated activities and the direct pressure it places on often very limited supply sources (Law, 2005).

In terms of the barriers presented above, it is foreseen that by far the most important emphasis will be placed on managing health risks to a level where public acceptance can be obtained. These two interlinked obstacles are expected to remain at the forefront of emerging DPR schemes, together with development of technological advances seen to present publicly acceptable, technologically robust and economically sound solutions. As the pressure increases, the focus will intensify on existing reuse projects where long-term effects have been extensively monitored and studied in an effort to contain uncertainties about the future and appease public fears.

For Windhoek, the future is unmistakably tied to intensified water reuse, to the point where the motto of 'every drop counts' becomes a reality to each and every citizen. Building on the success attained by past generations, the semi-purified supply network is currently being extended, to lighten even further the demand on potable water, while the planning of an additional direct potable reclamation facility is set to start towards the end of 2015.

Though the barriers presented herein are undeniably in existence, these are expected to come under severe assault in the coming decade as water reuse morphs from an attractive, environmentally friendly option to a forced, demand-driven reality.

Disclosure statement

No potential conflict of interest was reported by the author.

References

Bolton, S. (2014). Why understanding the true value of water is smart business [Blog]. Retrieved from http://www.greenbiz.com

Boucher, M., Jackson, T., Mendoza, I., & Snyder, K. (2011). *Public perception of Windhoek's drinking water and its sustainable future (Submitted in partial fulfillment of the requirements for the Degree of Bachelor of Science)*. Massachusetts, USA: Worcester Polytechnic Institute.

Brand, J.G. (1962). *Gammams Rioolsuiweringswerke*. Windhoek, Namibia: City of Windhoek – Department of the City Engineer

Cain, C.R. (2011). *An analysis of direct potable water reuse acceptance in the united states: Obstacles and opportunities* (Report No. 123). Baltimore, MD: Johns Hopkins Bloomberg School of Public Health.

Daigger, G.T. Lozier, J.C., & Crawford, G.V., (2006). Water reuse applications using membrane technology, Water Environment Foundation.

Department of Water Affairs and Forestry (DWAF). (2012). *Code of Practice Volume 6: Water Reuse*. Windhoek, Namibia: Ministry of Agriculture, Water and Forestry.

Du Pisani, P.L. (2004). Surviving in an arid land: Direct reclamation of potable water at Windhoek's Goreangab Reclamation Plant. *Arid Lands Newsletter*, 56 November/December 2004. Retrieved from http://ag.arizona.edu

European Commission – Environment. (2013). Background document to the public consultation on policy options to optimise water reuse in the EU [Consultation Report]. Retrieved from http://www.ec.europa.eu

FMG Goreangab Joint Venture. (1998a). *Goreangab Water Reclamation Plant (Design Report)*. Stuttgart, Germany: FMG Goreangab Joint Venture.

FMG Goreangab Joint Venture. (1998b). *Goreangab Water Reclamation Plant (Process Train Report)*. Stuttgart, Germany: FMG Goreangab Joint Venture.

Khalil, Z.A. (N/A) Membrane technology advances wastewater treatment and water reuse. Retrieved from http://www.cdmsmith.com

Law, I. (2005, June 10) Potable reuse: What are we afraid of? Potable Reuse. Retrieved from http://www.researchgate.net

Lee, H. & Tan, T. P. (2016). Singapore's experience with reclaimed water - NEWater. *International Journal of Water Resources Development*. doi: 10.1080/07900627.2015.1120188

Menge, J. (2006). *Treatment of wastewater for re-use in the drinking water system of Windhoek*. Windhoek: City of Windhoek/WISA.

Namibia Statistics Agency (NSA). (2013). Namibia 2011 population and housing census main report [Report]. Retrieved from http://nsa.org.na/page/publications

Paul, D. (2012). Value of Water [Blog]. Retrieved from http://www.growingblue.com

Sanz, L.A., & Gawlik, B.M. (2014). *Water reuse in Europe: Relevant guidelines, needs for and barriers to innovation*. Luxemburg: European Union.

UNDP. (2006). Human development report 2006. Retrieved from http://www.un.org/waterforlifedecade/scarcity.shtml

UNESCAP. (2012). Reusing and recycling water [Fact Sheet]. Retrieved from http://www.unescap.org

UNU-INWEH (2013). Rising Reuse of Wastewater in Forecast but World Lacks Data on "Massive Potential Resource" [Media Release]. Retrieved from http://www.unu.edu

USEPA. (2004). Guidelines for water reuse (Report No. EPA/625/R-04/108). Washington DC: U.S Environmental Protection Agency

Van der Walt, C. (2003). Multiple barriers ensure safe potable water from reclaimed sewage – Windhoek Namibia. The Water Wheel, July/August 2003.

Williams, R. (1996). *Wastewater reuse in residential and commercial situations*. Brisbane: CMPS&F. Churchill Fellowship.

WWAP (United Nations World Water Assessment Programme). (2015). *The United Nations world water development report 2015: Water for a sustainable world*. Paris, UNESCO.

A lived-experience investigation of narratives: recycled drinking water

Leong Ching

Lee Kuan Yew School of Public Policy, Institute of Water Policy, National University of Singapore, Singapore

ABSTRACT

Recycled drinking water (RDW) represents a cost-effective and technologically reliable source of urban water. Yet it remains one of the least implemented solutions because of emotional and psychological difficulties – the human dimension of the 'yuck factor', which has been empirically identified as statistically significant. Researchers have therefore recently expanded water research in RDW to include the psychology of users. This study builds on this effort by using the lived-experience methodology for the first time on RDW. Investigating the case of Singapore, the method reveals an 'insider's view' of key stakeholders, and uncovers human-scale narratives and experiences within the discourses of technology, economics of water supply, and ecological realities.

Introduction

A 2012 United Nations report on water states: "While most cities would refrain from using treated waste water as a source of drinking water, this avenue is also available and has been implemented, for example, in water-scarce Singapore and the International Space Station, without ill effects." (UNESCO (United Nations Educational, Scientific & Cultural Organization), 2012). The UN has also recommended the strategy of recycling wastewater to water-stressed countries, citing the particular case of Singapore.

Implementing recycled drinking water (RDW) is a contentious policy. To date, only three countries have done so on a large scale: Singapore, Namibia (Windhoek) and the United States (Orange County, in California), although many others have tried pilot projects in urban areas (Rygaard, Albrechtsen, & Binning, 2009). The most commonly cited obstacle is what has been called the 'yuck factor'; in fact, this factor has been found to be the one of the few statistically significant factors in empirical studies (Po et al., 2005). Explorations of this human visceral yuck factor underline an important and well-recorded psychological fact – namely that it exists and has strong policy impact (Leong & Yu, 2010; Po, Kaercher, & Nancarrow, 2003) – but there has been relatively little written about the actual experience of drinking such water. The case of Singapore presents a *prima facie* counterfactual to the prevailing view that the psychological barrier is an immutable one, a 'social fact' with its own logic that cannot be overcome by science and reason. The Singapore story has been well documented insofar as implementation of water policies, including RDW, has been concerned (Tortajada, 2006; Tortajada, Joshi, & Biswas, 2013). But, aside from the traditional empirical investigations, there is a rising need for what may be called hydrological sociology, or investigating the human dimension of change.

This is not easy because of the challenges inherent in applying a quantitative analysis to such subjective factors as public perception, discourses and narratives. This article is a complement to current research on institutional studies of perceptions, informal institutions and institutional change. It builds on an earlier quantitative study that used a Q methodology, which allows the examination of subjective viewpoints and perceptions, in the case of Singapore in implementing its water reuse policy. Using the lived-experiences method, it shows how the different discourses interact with and reinforce each other, to inform the formation of norms in water-reuse policies. Overall, the aim of this article is to illustrate the complexities of water-reuse implementation, locating it within research on informal institutions. It is the first study to be done on the lived experience of RDW.

The narrative of recycled water

The narrative of RDW is sometimes conceived as one of reason versus emotions – what we call a "thin" description (Lejano & Leong, 2012), which may not capture the complexities of the issue at hand, including the legitimate fear of health risks. Studies generally show that recycled water is safe for drinking. Tests have been carried out in Windhoek, Namibia (Hattingh, 1977) (Isaäcson & Sayed, 1988), Australia (Rodriguez et al., 2012) and the United States – in Florida, California and Arizona (Huffman, Gennaccaro, Berg, Batzer, & Widmer, 2006). Despite these findings, however, studies have pinpointed possible sources of risks, including microbial waterborne disease (Leder, Sinclair, & O'Toole, 2007), infectious microorganisms (Huffman, Gennaccaro, Berg, Batzer, & Widmer, 2006) and volatile organic compounds (Rodriguez et al., 2012). It is clear that health-effects research is important, as recycled water will be increasingly used in the urban and domestic context (Leder, Sinclair, & O'Toole, 2007). The yuck factor, however, appears to be quite different from the fear of these health risks.

This has been variously defined as a "psychological repugnance", "disgust", or "profound discomfort" (Marks, Martin, & Zadoroznyj, 2008). This is often contrasted with the apparently 'irrational' rejection of drinking recycled water. Framing the yuck factor in this manner is, however, a rather thin narrative – while it may be tempting to say that there is no rational basis for the yuck factor, it is nonetheless very easily rationalized (Stenekes, Colebateh, Waite, & Ashbolt, 2006), as seen from the discussion on health risks above.

Under the thin narrative of visceral reactions pitted against science and rationality, the explanation for the failure of water recycling schemes is that drinking recycled water is an activity supported by science but rejected by the public because of psychological reasons. These dominant ideas underline much of the effort of water managers in testing, quantifying and measuring public acceptance of water-reuse policies.

The pair of convictions is presented below:

H1: Irrationality Hypothesis. People don't accept RDW because of irrational fears. (Conversely, people accept RDW because they are able to overcome irrational fears by use of reason.)

H2: More Information Hypothesis. People don't understand the science and need more information. (Conversely, people can understand and were persuaded by scientific arguments.)

Because this article works on the reasons for acceptance, the converse framings of the two hypotheses were also included. This article aims to challenge these two hypotheses via an investigation of the narratives and discourses present. To do so, it undertakes a discursive institutional analysis of Singapore's case, which allows us to construct a narrative architecture of what it means to drink recycled water. We can then see how these confirm or falsify the two hypotheses outlined above.

A brief discourse on the methodological premise of using narratives to investigate informal institutions may be useful here. North (1990) gives probably the best-known definition of institutions as "the humanly devised constraints that structure human interaction. They are made up of formal constraints (rules, laws, constitutions), informal constraints (norms of behaviour, conventions, and self-imposed codes of conduct), and their enforcement characteristics."

Relating specifically to our enterprise of viewing narrative as part of these informal institutions, North (1986) provides us with a methodological premise when he suggests that institutions are "mental constructs" or "subjective models". His research leads him to investigations of learning and cognition. Within water institutions, Saleth and Dinar (2004, p. 26) see institutions as "subjective constructs". As Saleth and Dinar point out, it is not information per se but the *perception* of information that accounts for how institutions are constructed and changed (p. 64).

In this article, informal institutions are conceived of as forms of subjective constructs as manifested in public discourse and narratives. How then can we capture public narratives? We work off Crozier's (2007) idea of "recursive governance", where "the capacity for action, whether of governments, businesses or other societal actors, depends on their ability to engage in and manage open informational loops". More importantly, it is in such loops that "power and knowledge" are generated. Narratives then can be conceived of as "codified knowledge". Narratives are therefore a form of subjective construct, part of the informal institutions surrounding water. In our analysis, we take the perception (whether positive or negative) of water reuse as the object of investigation. The key question is: In one of the few places in the world where RDW has been implemented, what do the people themselves think of their experience, and what is their perception of water?

The place of water in Singapore

Singapore has been importing water since 1927. In 1961, this was formalized when the City Council signed the 1961 Water Agreement with the state of Johor in Malaysia. Under this agreement, Singapore had the "full and exclusive right and liberty to take, impound and use all the water" within the Gunong Pulai and Pontian catchments and Tebrau and Scudai Rivers up to 2011. In 1962, another agreement was signed "for the supply of up to 250 million gallons of water per day (mgd) from the Johor River, until 2061" (Tan, Lee, & Tan, 2009, pp. 139–140).

Singapore achieved full internal self-government (independent of the British) in 1959, and became part of Malaysia in 1963. Two years later, it was clear that the political merger of Singapore with Malaysia had failed. There was a deep divide, because of differences in beliefs over racial equality. This failed merger led to difficult bilateral relations, especially over the resource essential to life: water (Tortajada, 2006). At the time, there were about 1.9 million people (The Straits Times, 1966) living on the island, mainly crowded into the city centres. Public health provision was poor, and water-borne diseases such as cholera were common because of poor sanitation facilities. During the wet season, many parts of the city were underwater, whereas during the dry months, water had to be rationed.

Singapore has been experimenting with recycled water since 1974. Singapore's first pilot water reclamation plant, with a capacity of 381,360 litres of water per day, had problems, such as a strong smell of ammonia. It was subsequently shut down in late 1975 after the trial, and never released to the public (Channel News Asia, 2002; The New Paper, 2002).

By the 1990s, there were better-quality membranes, and the cost had halved (TODAY, 2002). It was around that time, soon after the start of the Asian Financial Crisis in 1997–1998, that relations between Singapore and Malaysia begun to sour. In 1999, three meetings were held between Singapore and Malaysia at the top official level to try to make progress on water and other bilateral issues (Tortajada, 2006). In the early 2000s, Malaysia itself suffered water shortages, and some quarters in Malaysia argued that water should go to Malaysians rather than being sold to Singapore.

Government officials revisited the idea of drinking recycled water (Public Utilities Board, n.d), going to the United States, including Orange County in Southern California, to study recycling methods (Channel News Asia, 2014). After the visit, the government constructed a demonstration plant to test recycled water. By May 2000, a SGD6.5 million plant had started operations (The New Paper, 2002). By 2001, the Public Utilities Board was releasing recycled water for non-potable use: wafer fabrication processes and non-potable applications in manufacturing processes, as well as cooling towers in commercial buildings. In September 2002, the name NEWater was given to RDW, an additional source of drinking water (Public Utilities Board, n.d). The next year, the Public Utilities Board introduced NEWater (at about 1% of total daily water consumption) into its water reservoirs. The amount was increased progressively to about 2.5% of total daily water consumption by 2011.

Methodology: lived experiences

The use of lived experience as a research methodology examines the everyday lived experience of human beings to provide concrete insights into the qualitative meanings of phenomena in people's everyday lives (Van Manen, 1997). In this article, the lived-experiences methodology builds on previous research in Singapore, which locates the narratives of RDW within key discourses (Leong, 2015). The supplementing of these broad narratives with the insider's view is useful for three reasons. First, its explicitly phenomenological approach uncovers different perceptions of RDW (Lindseth & Norberg, 2004), and by so doing aids in "answering questions of meaning" (Mapp, 2008, p. 308). This is the 'insider's view'. The data used are insights from a self-referent perspective, which can be gained only by the participants detailing their lived experience of a particular time in their lives (Clark, 2000). In the current investigation, the participants offer a privileged insider's view of their own body – "where the body can be understood or interpreted as being part of, and contributing to, a certain phenomenon" (Gill & Liamputtong, 2009) – particularly salient in drinking-water studies as water is tied so closely to health and life.

Second, this methodology is not limited to exploring the experiences of everyday lives but includes dealing with and responding to these experiences (Boylorn, 2008). This results in data that are "well-grounded, [with] thick and rich descriptions and explanations of processes situated within particular contexts" (Gill & Liamputtong, 2009, p. 314). Conceptually, the method locates itself within the larger scholarship of interpretative phenomenological analysis (IPA) (Smith, 1996, Eatough & Smith, 2010), a qualitative method that adopts for the study both an insider's perspective and the researcher's own viewpoint for interpreting the subject's perceptions. This dual orientation allows IPA to recognize both the power and the limits of context and language (Larkin & Thompson, 2012).

Third, lived experience is useful for its ability to connect everyday experiences with larger issues (de Casterlé et al., 2011), and in the process create "a space for storytelling interpretation, and meaning-making" within rigorous academic and quantitative research (Boylorn, 2008, p. 490). Although it has not been widely used in water studies, this methodology is increasingly popular in social science research (Gill & Liamputtong, 2009), for example in nursing (Phillips, 1993), medicine (de Casterlé et al., 2011; Finlay, 2011) and caring-sciences research (Lindseth & Norberg, 2004). The wide application of this method demonstrates its relevance and rigour in providing a good framework for understanding the complexity of the human dimension in social sciences. IPA is peculiarly suited for studying RDW because it provides a unique way of theorizing and investigating the connections between scientific understandings and subjective lived experiences – including that of the physical body. The body is at once a site and a means through which these everyday interactions occur (Finlay, 2011). The investigation relies on water as a commodity essential to life, contrasting the 'everydayness' of water in human lives (Sofoulis, 2005; Truelove, 2011) with the high technology of reverse osmosis and membranes. The methodology therefore provides an opportunity not only to study broader human experiences in relation to water but also to examine routine physical connections, and uncover the discourses that are formed as a result (Truelove, 2011).

Institutional change: personal narratives and public norms

The lived-experiences method outlines these discourses by showing how RDW is understood by different Singaporeans in the text below. Some 25 interviews were conducted within a span of four months, each interview lasting an average of 30 minutes. The longest interview lasted an hour, while the shortest interview was just 20 minutes. Respondents included men and women, with various age and racial groups. Semi-structured interviews were used to extract lived-experience responses, which were then analyzed using IPA (Eatough & Smith, 2010), which requires "open research questions, which focuses on the experiences, and/ or understandings, of particular people in a particular context" (Larkin & Thompson, 2012, p. 103). Three steps are necessary (Smith, Jarman, & Osborn 1999): (1) rereading of interview transcripts and identification of potential themes; (2) another reading of interview transcripts to identify themes, which are then organized; (3) defining themes in more detail and establishing the interconnections between themes (Smith, 1996; Lindsay, MacGregor, & Fry 2014). Therefore, IPA can be used to decode and make sense of responses in relation to the lived experiences of respondents.

Table 1. Narratives of NEWater from the Q methodology (Leong, 2015).

Factor	Ideational element
Factor 1	Technology can overcome water shortages
Factor 2	Water security remains a real problem for Singapore
Factor 3	Water should be priced to reflect the cost of supplying it
Factor 4	Innovations in water management, such as pricing and recycling, help ensure good supply of water
Factor 5	Singapore has severe physical and energy constraints and needs to choose the most cost-efficient way of producing water
Factor 6	Recycling water is a way of preserving independence and continued growth
Factor 7	The science of recycling still needs to be better known, and there is still an instinctive rejection of drinking sewage water ('yuck factor')
Factor 8	As climate change and global water scarcity take root, recycled drinking water is a way to ensure water supply

In this article, we modify this approach using the results from an earlier discourse analysis (Leong, 2015), which outlined three key discourses on technology (DIscourse 1), the economics of water (Discourse 2) and the environment (Discourse 3). The seven factors relied on in the discourse below are listed in Table 1. The discourses are then illustrated with the actual quotes and lived experiences of the interviewees (Table 2).

Discourse 1: Technology can change current paradigms

- Factor 1: Technology can overcome water shortages.
- Factor 2: Innovations in water management, such as pricing and recycling, help ensure good supply of water.

Discourse 2: Water security is a problem with an economic cost

- Factor 3: Water security remains a real problem for Singapore.
- Factor 4: Water should be priced to reflect the cost of supplying it.
- Factor 5: Singapore has severe physical and energy constraints and needs to choose the most cost-efficient way of producing water.

Discourse 3: Environmental and global realities make it imperative to recycle water

- Factor 6: Recycling water is a way of preserving independence and continued growth.
- Factor 7: As climate change and global water scarcity take root, recycled drinking water is a way to ensure water supply.
- Factor 8: The science of recycling still needs to be better known, and there is still an instinctive rejection of drinking sewage water.

In an earlier quantitative analysis using the Q methodology, three key discourses were identified, derived from eight factors (Table 1). These key discourses showed that the broad narratives of RDW in Singapore were clustered along three lines: economics of water, change

Table 2. Quotes and lived experiences of the interviewees.

Primary themes	Sample text
Discourse 1: Technology can change current paradigms • Factor 1: Technology can overcome water shortages • Factor 2: Innovations in water management, such as pricing and recycling, help ensure good supply of water	Participant 1 (P1): I trust our water supply a lot. I always drink it from the tap. I know that the water from our tap is relatively safe for consumption. That is something that I am very thankful for, because if we compare to other less developed countries, it is not very easy for them to get cleaned water supply, even if you go there and you do this water purification thing for them. I have a friend who went there to do [this]. They set up the portable water purification. I don't know what happened, but at the end my friend said it got contaminated, which meant the people cannot drink the water anymore (P1) I think Singapore has enough water because we are looking into a way that meant to make sure that we have the technology, so even if we don't have the fresh water, we can have the recycled one (P7) I would say it's really based on faith, going by how the government is planning for it in terms of the technology they have put in place to ensure we have a steady flow of water. I say I have probably 75 per cent confidence that we will have enough water for the future, and in the event of an adverse condition. The balance 25 per cent [*sic*] is really about the climate. I mean our usage is a constant thing, we can foresee how our consumption will change but it's really the supply. Because supply is really affected by rain. Climatic events (P2) I don't think we will run out of water because we have the NEWater plants and desalination plants. Seawater is everywhere – unless the sea dries up, I don't think we will run out of water (P3) Politically, I don't think more expensive [water] would fit very well for people, but for me I would not mind paying more, because I think that would challenge people not to waste water. I think people don't really realize the need for water all the time. I don't know if a campaign to save water would work in Singapore. I personally feel that education is important, but if people don't feel it [now], it is not going to matter to them later
Discourse 2: Water security is a problem with an economic cost • Factor 3: Water security remains a real problem of Singapore • Factor 4: Water should be priced to reflect the cost of supplying it • Factor 5: Singapore has severe physical and energy constraints and needs to choose the most cost-efficient way of producing water	(P2) They can just cut off anytime. I rather take from NEWater rather than dependent on Malaysia (P4) I try not to think about it. In Singapore, any source of water is good. As long it is potable (P1) My idea about water has been shaped quite a lot by my overseas experience, because when I go overseas, it becomes like 'we have a limited supply'. Where there is limited supply, we know you have to treasure what you have. But for Singaporeans, it is there … on the tap and in public areas also. So they don't feel the need to conserve. But when I go overseas, I see how the children have to drink even polluted water (P6) To a large extent, I think Singapore is ready for floods. Droughts I'm not too sure, because we are importing water – so I guess if Malaysia doesn't threaten us politically we should be fine. We are working towards self-sufficiency. There should not be any issues about droughts. Flooding, it should be fine. I trust the government quite a lot. Our buildings are all high-rise buildings, so that shouldn't be an issue. Our drainage is really good, too (P8) It's a necessity, we just don't have enough. And I think they are doing what they can to increase the source through various means – like seawater, filtration, NEWater and all that stuff. But until we are fully dependent on our water source, we still need it. Otherwise, our industries and all cannot run (P6) I guess that comes along with, like when I initially came back from Nepal, I was very careful with how much water I used. After a while, you just … When I came back from Nepal, I turned on the tap to wash my hands, it was very little. After a while, I just turned it on causally, wash it and go – I don't care. I guess in my mind, the only consequence is higher water bills, it doesn't directly relate to less water for others – or like you are wasting water. It's not an environmental consequence, it's more of a financial consequence – like if I waste water, I have higher bills. In that sense, we might not want to waste water. But I don't think it's an environmental consequence for us (P5) I don't think it will run out. It's there, because we are surrounded by sea. So we've got plenty of water. So when the need arises, there will always be a way to make it usable. So I don't think we will run out of water in that sense, but the question is, how much will it cost? And that's a tangible thing

(Continued)

Table 2 *(Continued).*

Primary themes	Sample text
	(P7) Water doesn't feel like it's in shortage. We don't have rationing and stuff like that, it's just in constant supply so I guess people just don't think about it in terms of the waste
	(P1) I don't think many countries would have the NEWater, because the technology behind NEWater is very expensive. And I think we lack natural resources, more than other countries, they have large water bodies. Water is not their main primary concern, but for us it becomes one of our largest concerns
	(P7) I guess there is a certain level of complacency as we have not seen water shortage, as in we have not seen the effects of it. The new technologies, desalination and NEWater, give us both complacency and a false sense of security, whichever. But I know in the future, it's more a question about cost. Desalination, from what I know about it, the basic thing about it is the cost, and technology is expensive. If it becomes a main source of water, our water bills will go up by a lot for sure
Discourse 3: Environmental and global realities make it imperative to recycle water • Factor 6: Recycling water is a way of preserving independence and continued growth • Factor 7: As climate change and global water security take root, recycled drinking water is a way to ensure water supply • Factor 8: The science of recycling still needs to be better known, and there is still an instinctive rejection of drinking sewage water	(P2) Most people fear that we are dependent on Malaysia, but they do not make an effort to save water. NEWater is everywhere and it is more prominent than Malaysia's water. So people think 'Never mind, I got the NEWater.' The feeling is that we are confident we can recycle our own water, so we are not so dependent on Malaysia's anymore
	(P5) Reliance and independence. It is better to be self-sufficient, as much as possible. Singapore is already dependent in many ways on other countries
	(P4) I think recycled water definitely helps. I mean most of it goes to the industries so I don't think it's a big concern for drinking supply. It definitely supplements our drinking water supply
	(P2) My parents drink water straight from the tap. Only I in the family boil the water. The rest of my siblings just drink water from the tap, because it is easy. Because I went to the NEWater plant, and I found out that NEWater is actually the water from toilet water and stuff. And then, I say I should just boil water from now on
	(P2) Because I have seen people's water bills are 1 dollar plus only, I am not that much frugal user of water. I think the older generation like my grandma makes an effort to save water.… My grandma told us, 'Don't bathe too long' or 'Don't leave the tap running when you wash the dishes.' They listen to the radio, people talking about saving water
	(P2) I don't pay the water bill, so I don't really know if the water is expensive in Singapore. The only thing I make effort to save was when I wash the dishes, I don't use too much water. However, when I take a shower, I want all the water
	(P2) I think most of Singaporeans are wasteful. When you have the money, you really can afford all the water bills. So you don't really have to save that much. Unless you have low income, that you have to save
	(P2) When I was very young, we were introduced to NEWater and it was very simply put as water from the toilet that was cleaned into drinking water. I was pretty hesitant about it of course. I was like, 'Hey, I'm not going to drink that water.' But when we were given the water, I tasted it and it tasted a bit weird.… Maybe they tried too hard to purify the water or something. After that when I understood the processes involved when I did my studies, I knew how the water was being cleaned, I realized how NEWater is extremely clean. Even cleaner than normal water I guess, the water which we export. Then I realized it's perfectly fine to drink it. But I still think it tastes a bit weird, I think it tastes somewhat weird
	(P3) I have a passion for sustainability. My family is a frugal water user. I think of when I was younger. My family tried to tell us not to waste water within the family, save money, save water. When I grew older, I realized water does not cost that much for the household. But it is more of a personal sustainability view that I take on. My attitude toward water was part of the attitude toward the environment. I don't buy bottled water. I bring my own water bottle, because I hate creating unnecessary waste. I guess the concept about NEWater was a bit weird when I first learned about it at younger age. For me, as long as it is safe and it is certified as safe, I would just drink it anyway. It is not a big deal to me. By certify, I refer to the government. I trust in government's 'certify'

in paradigms from use of technology, and geopolitical realities (Leong, 2015). An analysis with the lived-experience methodology gives us two further insights:

1. The three discourses are not separate but mutually reinforcing. This we illustrate with the case of P2 (the same person citing different lived experiences that speak to two discourses), as well as an exploration of the issue of trust, which also ranges across discourses.

2. Together, these discourses form a coherent and compelling narrative which speaks against H1 and H2, sometimes singly, sometimes in unison. The notion of 'rationality', therefore, is not just a matter of reason and science but also that of meaning and coherence with a larger social reality; that is, rationality with regard to decision making in RDW can be seen as an interpretative concept that allows people to make sense of their social reality.

Looking at Discourse 2 in particular, both H1 and H2 appear to be contradicted by the lived experience of the participants on the economics of water. This discourse overall is well established in Singapore. It became a salient issue when Singapore and Malaysia were locked in intense discussions over the price of water. When Singapore adopted RDW, the pricing of NEWater was a subject that was openly discussed, together with the price of alternatives such as desalination. Singapore's general model for water pricing was a cost-recovery one, a principle that had been established before the introduction of NEWater (Tortajada, 2006). After NEWater, when it was mixed into the general water supply system, the government made the difficult decision for all end users to pay NEWater cost as the marginal cost of water (Tortajada, Joshi, & Biswas, 2013). NEWater was a cheaper option than desalination (Tortajada, 2006), but of course more costly than rainwater. This economic frame of water had two large impacts. First, recycled water was merely one option, to be weighed against others in terms of economic cost – it was not *sui generis*. Second, the building of large and expensive desalination plants underlined the economic valuation of water, alongside the security and strategic dimensions. This is well captured by the experiences below:

(P2) Because I have seen people's water bills are 1 dollar plus only, I am not that much frugal user of water. I think the older generation like my grandma makes effort to save water.... My grandma told us, 'Don't bathe too long' or 'Don't leave the tap running when you wash the dishes.' They listen to the radio, people talking about saving water.

(P2) I think most of Singaporeans are wasteful. When you have the money, you really can afford all the water bills. So you don't really have to save that much. Unless you have low income, that you have to save.

(P1) I don't think many countries would have the NEWater, because the technology behind NEWater is very expensive. And I think we lack natural resources, more than other countries, they have large water bodies. Water is not their main primary concern, but for us it becomes one of our largest concerns.

The lived experiences within Discourse 2 also contradict H2

Neither P2 nor P3, who was more explicitly supportive, made reference to the science of NEWater, and the fact that the scientific community persuaded them. A look at the explicitly technological strand of the discourse (Discourse 3) explains how the role of technology is perceived. First, it is thought of as a way towards water self-sufficiency (P1: 'I think Singapore has enough water because we are looking into a way that meant to make sure that we have the technology, so even if we don't have the fresh water, we can have the recycled one'). There is also a link with the economics of water – linking the price of water with what is perceived to be a high-technology way of reclaiming water:

> (P5) I don't think many countries would have the NEWater, because the technology behind NEWater is very expensive. And I think we lack natural resources, more than other countries, they have large water bodies. Water is not their main primary concern, but for us it becomes one of our largest concerns.

The lived experiences of P2 challenge H2 as well. While she does drink recycled water, she boils it first, despite knowing (or being in possession of the fact that) the quality of the water is high enough to drink:

> Only I in the family boil the water. The rest of my siblings just drink water from the tap, because it is easy. Because I went to the NEWater plant, and I found out that NEWater is actually the water from toilet water and stuff. And then, I say I should just boil water from now on.

Another participant, P3, also shows the limitations of casting public narratives in this binary fashion. Rather than a matter of reason trumping emotions, it was trust in the authorities that enabled her to do so:

> The concept was a bit weird when I first learnt about it.… For me, as long as it is safe and it is certified as safe, I would just drink it anyway. It is not a big deal to me. By certify, I refer to the government. I trust in government's 'certify'.

The issue of trust ranges across the different discourses, perhaps unsurprisingly since empirical studies have shown that public trust in regulatory actors and stakeholders determines people's attitudes and actions (Priest, Bonfadelli, & Rusanen, 2003; Brunk, 2006). This is especially true in the area of science and technology (de Jonge, van Trijp, van der Lans, Renes, & Frewer, 2008). When these institutions are not perceived as legitimate or credible, policy or institutional change is likely to fail, a finding upheld in policy changes as wide-ranging as the introduction of new technologies (Grove-White, Macnaghten, Mayer, & Wynne, 1997, Siegrist & Cvetkovich, 2000, Durant & Legge, 2005), biotechnology in the United States, (Priest, 2001), and support for gene therapy, human cloning and genetic databases in Britain (Barnett, Cooper, & Senior, 2007). In our investigation, trust is a key part of the lived experience, and appears in Discourses 1 and 3. P3 speaks about trust but she also chose drinking tap over bottled water because of the environment narrative: 'My attitude toward water was part of the attitude toward environment. I don't buy bottled water. I bring my own water bottle, because I hate creating unnecessary waste.' With the sense of imperative emanating from a long-term event such as climate change, the need to recycle water becomes part of larger narrative of sustainability.

Trust was also present in Discourse 1 (P1):

> I trust our water supply a lot. I always drink it from the tap. I know that the water from our tap is relatively safe for consumption. That is something that I am very thankful for, because if we compare to other less developed countries, it is not very easy for them to get cleaned water supply, even if you go there and you do this water purification thing for them. I have a friend who went there to do [this]. They set up the portable water purification. I don't know what happened, but at the end my friend said it got contaminated, which meant the people cannot drink the water anymore.

This extract shows that there was a trust, not just in the technology – because there is evidence that the technology of purification per se is not fool-proof, as evidenced by the experience of the portable device in Cambodia. Rather, technology is located within a larger system, which together, for this participant, engenders trust.

Conclusion

This article represents the first effort at using the lived-experience method to uncover the attitudes and perceptions of people who experience and drink recycled water as part of their daily lives. RDW has often been considered part of social norms – the informal institutions surrounding water issues. We peer into this analysis and uncover what this means from an insider's point of view – that is to say, within the larger societal discourses, that are the experiences and beliefs of the people who live with this phenomenon as part of their daily lives.

The article has three important implications. First, it requires us to rethink the two dominant hypotheses which are aligned with RDW at the moment – irrationality versus science, and the need for more information. Both are falsified by the lived experiences of the participants.

Second, the article also outlined how the discourse of water-reuse polices was located within larger, mutually reinforcing narratives such as the economics of water and the rising ecological limits of water supply and use. The basic narrative in RDW has therefore been transformed from a purely 'norms-based' approach, focusing on the prevailing psychological reaction to drinking recycled water, to a more complex one, incorporating economic interests as well as national security. The narrative was no longer an either/or dichotomous discourse that pits the pro- against the anti-water-reuse camp – a thin narrative – but one that is thicker, making room for both camps, but allowing a principled stance for the acceptance of recycled water.

Third, in terms of research approaches, it shows the importance of a multi-layered understanding of RDW, one which does not appear to be completely captured by the usual quantitative surveys that test public acceptance of RDW. Stripped of larger political and social imperatives, the logic of norms, discourses and narratives cannot take root. Yet it is these often ignored variables that are required if informal institutions are to have a place in illuminating the dynamics of institutional change.

Acknowledgment

The author thanks Mr Thinesh Kumar for his invaluable research assistance.

Disclosure statement

No potential conflict of interest was reported by the author.

References

Barnett, J., Cooper, H., & Senior, V. (2007). Belief in public efficacy, trust, and attitudes toward modern genetic science. *Risk Analysis, 27*(4), 921–933.

Boylorn, R. M. (2008). As seen on TV: An autoethnographic reflection on race and reality television. *Critical Studies in Media Communication, 25*, 413–433.

Brunk, C. G. (2006). Public knowledge, public trust: Understanding the 'knowledge deficit'. *Community Genetics, 9*(3), 178–183.

Channel News Asia. (2002, August 28). *Thirsting for new sources*. Singapore: Mediacorp.

Channel News Asia. (2014, March 24). *CALIFORNIA: Singapore looks to California for lessons on water management* [*News clip*]. Singapore: Mediacorp.

Chen, Z., Ngo, H. H., & Guo, W. S. (2012). 2013;). Risk control in recycled water schemes. *Critical Reviews in Environmental Science and Technology, 43*, 121127060944004. doi: 10.1080/10643389.2012.672085.

Clark, J. (2000). Beyond empathy: An ethnographic approach to cross-cultural social work practice. *Unpublished manuscript, Faculty of Social Work, University of Toronto.*

de Casterlé, B. D., Verhaeghe, S. T., Kars, M. C., Coolbrandt, A., Stevens, M., Stubbe, M. & Grypdonck, M. (2011). Researching lived experience in health care: Significance for care ethics. *Nursing Ethics, 18*, 232–242.

de Jonge, J., van Trijp, J. C. M., van der Lans, I. A., Renes, R. J., & Frewer, L. J. (2008). How trust in institutions and organizations builds general consumer confidence in the safety of food: A decomposition of effects. *Appetite, 51*(2), 311–317.

Durant, R. F. & Legge, J. S. (2005). Public opinion, risk perceptions, and genetically modified food regulatory policy: Reassessing the calculus of dissent among European citizens. *European Union Politics, 6*, 181–200.

Eatough, V. & Smith, J. A. (2010). Interpretative phenomenological analysis. In C. Willig & W. Stainton-Rogers (Eds.), *The SAGE handbook of qualitative research in psychology* (pp. 179–195). London, UK: SAGE Publications.

Finlay, L. (2011). *Phenomenology for therapists*. NJ, USA: John Wiley & Sons.

Gill, J. & Liamputtong, P. (2009). 'Walk a mile in my shoes': Researching the lived experience of mothers of children with autism. *Journal of Family Studies, 15*, 309–319.

Grove-White, R., Macnaghten, P., Mayer, S., & Wynne, B. (1997). *Uncertain world: genetically modified organisms, food and public attitudes in Britain*. Lancaster, UK: Centre for the Study of Environmental Change, Lancaster University.

Hattingh, W. (1977). Reclaimed water: A health hazard? *Water S. A, 3*, 104-112.

Huffman, D. E., Gennaccaro, A. L., Berg, T. L., Batzer, G., & Widmer, G. (2006). Detection of infectious parasites in reclaimed water. *Water Environment Research, 78*(12), 2297–2302. Retrieved from http://www.jstor.org.libproxy1.nus.edu.sg/stable/25053635

Isaäcson, M. & Sayed, A. R. (1988). Health aspects of the use of recycled water in Windhoek, SWA/Namibia, 1974-1983. diarrhoeal diseases and the consumption of reclaimed water. *South African Medical Journal = Suid-Afrikaanse Tydskrif Vir Geneeskunde, 73*, 596.

Larkin, M. & Thompson, A. (2012). Interpretative phenomenological analysis. In A. Thompson & D. Harper (Eds.), *Qualitative research methods in mental health and psychotherapy: a guide for students and practitioners* (pp. 99–116). Oxford, UK: John Wiley.

Leder, K., Sinclair, M., & O'Toole, J. (2007). Recycled water and human health effects. *Australian Family Physician, 36*, 998–1000.

Lejano, R. P. & Leong, C. (2012). A hermeneutic approach to explaining and understanding public controversies. *Journal of Public Administration Research and Theory, 22*, 793–814.

Leong, C. (2015). A quantitative investigation of narratives: recycled drinking water. *Water Policy, uncorrected proof,* 1–17.

Leong, C. & Yu, D. (2010). Turning the tide: Informal institutional change in water reuse. *Water Policy, 12*, 121–134.

Lindsay, H., MacGregor, C., & Fry, M. (2014). The experience of living with chronic illness for the haemodialysis patient: An interpretative phenomenological analysis. *Health Sociology Review: The Journal of the Health Section of the Australian Sociological Association, 23*(3), 232–241.

Lindseth, A. & Norberg, A. (2004). A phenomenological hermeneutical method for researching lived experience. *Scandinavian Journal of Caring Sciences, 18*, 145–153.

Mapp, T. (2008). Understanding phenomenology: The lived experience. *British Journal of Midwifery, 16*, 308–311.

Marks, J., Martin, B., & Zadoroznyj, M. (2008). How Australians order acceptance of recycled water: National baseline data. *Journal of Sociology, 44*, 83–99.

Phillips, J. R. (1993). Researching the lived experience: Visions of the whole experience. *Nursing Science Quarterly, 6*, 166–167.

Po, M., Kaercher, J. D. ,& Nancarrow, B. E. (2003). *CSIRO Land and Water (Dec 2003) Technical Report 54/03,* 1–39.

Po, M., Nancarrow, B., Leviston, Z., Porter, N., Syme, G., & Kaercher, J. (2005). *Predicting community behaviour in relation to wasterwater reuse: What drivers decisions to accept or reject?.* Perth, Western Australia: CSIRO Land & Water.

Priest, S. H. (2001). Misplaced faith: Communication variables as predictors of encouragement for biotechnology development. *Science Communication, 23*(2), 97–110.

Priest, S. H., Bonfadelli, H., & Rusanen, M. (2003). The "Trust gap" hypothesis: Predicting support for biotechnology across national cultures as a function of trust in actors. *Risk Analysis, 23*(4), 751–766.

Rodriguez, C., Linge, K., Blair, P., Busetti, F., Devine, B., Van Buynder, P., & Cook, A. (2012). Recycled water: Potential health risks from volatile organic compounds and use of 1,4-dichlorobenzene as treatment performance indicator. *Water Research, 46,* 93–106. doi: 10.1016/j.watres.2011.10.032.

Rygaard, M., Albrechtsen, H. J., & Binning, P. J. (2009). *Alternative water management and self-sufficient water supplies.* London: IWA Publishing.

Saleth, R. M., & Dinar, A. (2004). *The institutional economics of water: a cross-country analysis of institutions and performance.* Cheltenham, UK: Edward Elgar Publishing.

Siegrist, M. & Cvetkovich, G. (2000). Perception of hazards: The role of social trust and knowledge. *Risk Analysis, 20*(5), 713–720.

Smith, J. A. (1996). Beyond the divide between cognition and discourse: Using interpretative phenomenological analysis in health psychology. *Psychology & Health, 11,* 261–271.

Smith, J. A., Jarman, M., & Osborn, M. (1999). Doing interpretative phenomenological analysis. In M. Murray & K. Chamberlain (Eds.), *Qualitative health psychology: theories and methods* (pp. 218–241). London: Sage.

Sofoulis, Z. (2005). Big water, everyday water: a sociotechnical perspective. *Continuum: Journal of Media & Cultural Studies, 19,* 445–463.

Stenekes, N., Colebateh, H. K., Waite, T. D., & Ashbolt, N. J. (2006). Risk and governance in water recycling: Public acceptance revisited. *Science, Technology and Human Values, 31,* 107–134.

Tan, Y. S., Lee, T. J., & Tan, K. (2009). *Clean, green and blue: Singapore's journey towards environmental and water sustainability.* Singapore: Institute of Southeast Asian Studies.

The New Paper. (2002, September 6). Nothing new about NEWater, really. *The New Paper,* p. 8.

The Straits Times (1966) 1.9 million Singaporeans. *The Straits Times,* 2 November 1966. Retrieved Febuary 10 2016 from: http://eresources.nlb.gov.sg/newspapers/Digitised/Article/straitstimes19661102.2.99.aspx

TODAY. (2002, September 6). NEWater an old idea. *TODAY Newspaper,* p. 9.

Tortajada, C. (2006). Water management in Singapore. *International Journal of Water Resources Development, 22,* 227–240.

Tortajada, C., Joshi, Y., & Biswas, A. K. (2013). *The Singapore water story: Sustainable development in an urban city state.* London: Routledge.

Truelove, Y. (2011). (Re-) Conceptualizing water inequality in Delhi, India through a feminist political ecology framework. *Geoforum, 42,* 143–152.

UNESCO (United Nations Educational, Scientific and Cultural Organization). (2012). *The United Nations World Water Development Report 4 (WWDR4).* Paris: UN-Water.

Van Manen, M. (1997). *Researching lived experience: Human science for an action sensitive pedagogy.* London: Althouse Press.

Public acceptance and perceptions of alternative water sources: a comparative study in nine locations

Anna Hurlimann[a] and Sara Dolnicar[b]

[a]Faculty of Architecture, Building and Planning, The University of Melbourne, Australia; [b]UQ Business School, The University of Queensland, Brisbane, Australia

ABSTRACT
Public acceptance of recycled water, desalinated water and rainwater is compared across nine international locations: Australia, Belgium, Canada, Israel, Japan, Jordan, Mexico, Norway and United States (specifically in Los Angeles). An online study was conducted in 2012, with 200 participants recruited to be representative of their respective location (1800 in total). The study investigated participants' intended use of and perceptions of alternative water sources. Results indicate that respondents clearly differentiate between alternative water sources. Water source preference varied between water use purposes. Significant differences were found between locations in the percentage of respondents willing to use alternative water sources. Additionally the study found significant differences across locations in perceptions held of five water sources.

Introduction

An ample supply of clean water is critical to sustain human life and the environments on which humans and nature depend. Humans have altered natural water regimes over time with the aim of securing an adequate supply of water to meet the needs and demands of population settlements (Mumford, 1989). For over a century, infrastructure-intensive, centralized water supply systems have dominated the provision of water in cities of developed nations. Yet, in many developing nations, the provision of water supply is not widespread, and when it does exist, is often unreliable in terms of quality and continuity of supply (World Water Assessment Programme (United Nations), 2009). To address quality concerns, consumers in such situations often diversify the sources of water they access and employ some type of water treatment – such as boiling water – to improve quality (Rosa & Clasen, 2010).

The sustainability of the centralized, supply-side approach to urban water supply management is being challenged in many locations due to the increasing pressure placed on these systems by changes in population, land use, environment and climate. Because of the limitations of traditional centralized supply systems, new approaches to water management have been called for. These include the introduction of the concept of "integrated water resources management" (Biswas, 2008) and augmentation of existing water supplies with nontraditional sources such as recycled wastewater and desalinated seawater.

There are numerous examples of the successful implementation of water augmentation projects, including many desalination plants, for example in the city of Perth, Australia, where 41% of water is sourced from desalination (Water Corporation, 2015). Successful potable recycled water projects include but are not limited to Singapore (Leong, 2015) and Namibia (du Pisani, 2005). However, the proposed solutions to current water supply challenges are not without hurdles. The implementation of alternative water sources for potable purposes has faced barriers in some instances, for reasons including poor political support and public acceptance – see e.g. the case of a proposed desalination plant in Sydney, Australia (Davies, 2006) and the proposed use of recycled water for potable purposes in Toowoomba, Australia (Hurlimann & Dolnicar, 2010). Further knowledge of the factors that contribute to the successful introduction of alternative water sources will provide critical information to meet future water needs under changing conditions.

Knowledge surrounding the factors contributing to the acceptance of alternative water sources is growing, thanks to an increasing body of research. While there is a substantial collection of research into recycled water acceptance, less detailed knowledge exists for other alternative water sources such as desalinated seawater, stormwater and rainwater. Additionally, attitudes across alternative water sources or locations have rarely been compared, and no large cross-nation study of public acceptance of alternative water sources has been published to date. Such research would provide insights into the similarities and differences across divergent water, cultural and social contexts. The key proposition of this article is that these differences in local water, political, environmental and cultural situations will be associated with differences in perceptions and acceptance of water from alternative sources. Hence, the research questions this article addresses are:

- Are there significant differences in stated willingness to use recycled water, desalinated sea water and rainwater between people in locations characterized by different water circumstances?
- Do perceptions of recycled water, desalinated seawater, rainwater, bottled water and current tapwater vary between people in locations characterized by different water circumstances?

This knowledge gap was addressed through a study in 2012 which simultaneously conducted surveys with 200 people in each of nine locations (1800 participants in total) in nine countries. Public willingness to use and perceptions of alternative water sources were compared across the nine locations. The article begins with an overview of previous research relating to willingness to use and perceptions of alternative water sources, before detailing the research method employed, and providing information about the locations of study. The results are then presented and discussed, and conclusions drawn.

Previous studies on acceptance of alternative water sources

Research into community acceptance of water alternatives has been conducted across a wide range of locations, including Europe (Aitken, Bell, Hills, & Rees, 2014; Jeffrey & Jefferson, 2003) and the Middle East (Alhumoud & Madzikanda, 2010; Al-Mashaqbeh, Ghrair, & Megdal, 2012; Carr, Potter, & Nortcliff, 2011). A significant body of research has developed in the

US since the 1970s focusing on recycled water acceptance (e.g. Bruvold, 1972, 1988, 1992; Bruvold & Ward, 1970; Haddad, Rozin, Nemeroff, & Slovic, 2009; Macpherson & Snyder, 2013; Ormerod & Scott, 2013; Resource Trends Inc., 2004). Similarly, a long history of research into public acceptance of recycled water use is evident in Australia (e.g. Fielding & Roiko, 2014; Hurlimann, 2008; Marks, 2004; Sydney Water, 1996). Despite this extensive research, knowledge gaps do exist, hampering a thorough understanding of the conditions which facilitate public acceptance of a range of alternative water sources and how acceptance can be influenced across locations. An overview of prior research in this field, and the gaps which exist, is provided below.

Understanding public acceptance of alternative water sources

Much work has been dedicated to the study of public perceptions of recycled water, the most extensively researched alternative water source. Most of this work was hypothetical in nature: the survey participants had never actually experienced recycled water. A key finding that emerged from this body of work is that public acceptance of recycled water is higher when the water does not come in direct contact with the body. For example, using recycled water to irrigate gardens is more acceptable than drinking it (e.g. Alhumoud & Madzikanda, 2010; Browning-Aiken, Ormerod, & Scott, 2011; Bruvold, 1988; Marks, Martin, & Zadoroznyj, 2006; McKay & Hurlimann, 2003; Sydney Water, 1996). The specific uses of recycled water which have been investigated have varied from study to study, as has the number of uses investigated, yet this overall acceptance trend prevails. It is important to note that the exact proportion of the public found to support the use of recycled water varies between studies and across time periods.

In Hurlimann's (2008) study of willingness to accept recycled water at four periods of time between 2002 and 2007, attitudes were not stable over the study period. The author believes that the differences in attitudes observed over these time periods could be explained by factors relating to the introduction of nonpotable source water for the recycled water system into the suburb, Mawson Lakes in South Australia. Similarly, Price et al.'s (2010) long-term study with residents of South East Queensland, conducted from November 2007 to December 2008, found that support for indirect potable reuse changed over time (decreasing significantly across two of the four time periods; $p < .01$ between the first two and last two time periods). The highest level of support (73.9%) was found during the baseline study. Support was at its lowest in the final survey period (70.2%). As Price et al. discuss, the surveys were conducted at a critical time for the region of South East Queensland in terms of water supply. A potable reuse scheme was being planned and constructed. During the first survey the region's dams were at 20% of capacity. By December 2008 the dams were back to 40% of capacity, and the government announced that the dams would not be augmented with recycled water unless their capacity decreased to a critical point. Price et al.'s (2010) additional findings indicate that many respondents would prefer not to drink potable recycled water and that their willingness to do so would decrease if other alternatives were available, including further rainfall.

In comparison, acceptance of desalinated water is not as extensively investigated. Theodori, Avalos, Burnett, and Veil (2011) surveyed 1228 residents of the state of Texas by mail in 2008 to gauge perceptions of whether desalinated water could be safely used for a range of purposes. They found acceptance patterns similar to those for recycled water. More recent research in Perth, Australia, compared public attitudes to using desalinated water for the city's supply (Gibson, Tapsuwan, Walker, & Randrema, 2015). The authors found that acceptance of desalination remained constant over the two survey periods (74% in 2007 and 73% in 2013). Psychological variables were the most dominant factors driving acceptance, including perceived outcomes, fairness, and low perception that the system may fail.

Recent studies have investigated attitudes to the use of rainwater. A number of studies have focused on factors facilitating increased use (e.g. Barthwal, Chandola-Barthwal, Goyal, Nirmani, & Awasthi, 2014; Domènech & Saurí, 2011; Gabe, Trowsdale, & Mistry, 2012). Dobrowksy et al. (2014) conducted 68 interviews with residents of the Kleinmond development in South Africa. All houses in the development have a rainwater tank, but at the time of interview did not have a municipal (centralized) water supply. The interviews addressed respondent use of rainwater, and found that 92% of respondents use the rainwater for laundry, 70% for cleaning, 46% for gardening, 44% for bathing, 24% for drinking and 19% for cooking. Of the 24% who indicate they use it for drinking, the majority indicated they only do so sometimes. The study did not investigate the principal source of water used for drinking, or whether the participants boil water for drinking.

Similarly, in the Mekong Delta region of Vietnam, Özdemir et al. (2011) surveyed 619 households to understand current practices and preferences regarding rainwater harvesting and other water sources. The study found that rainwater was the most frequently stated source of water used across both the wet and dry seasons (85% wet season, 84% dry season). While only 78% of respondents indicate they use rainwater for cooking, well water was used by 9% of respondents for drinking, and 17% for cooking. These findings point to people differentiating clearly between water sources and displaying different preferences.

Attitudinal differences between alternative water sources

A small number of studies have compared public attitudes to different water sources. Dolnicar and Schafer's (2009) Australian study compared stated preferences for recycled water and desalinated water across 20 water use purposes. Desalinated water was the preferred water source for close-to-body uses such as drinking and cooking. Recycled water was preferred for less personal uses such as watering the garden and flushing the toilet. Dolnicar and Hurlimann's (2010) later study, also conducted in Australia, found an overall stated preference for desalinated water over recycled water. Recycled water was preferred only for watering the garden. In a repeat cross-sectional study in Australia, Dolnicar, Hurlimann, and Grün (2014) asked respondents about their preferred drinking water source in January and July 2010. Ordered preference remained constant, with tapwater the preferred source (45%, 44%), followed by bottled water (28%, 27%), rainwater from their own tank (24%, 26%), desalinated water (1%, 3%), and finally recycled water (1%, 1%).

A research study conducted for the WateReuse Research Foundation in the US (Macpherson & Snyder, 2013) sought to investigate the impact of information on acceptance. Specifically, it investigated whether presenting recycled water use in the context of the urban water cycle, and the fact that all water is recycled, would increase acceptance of recycled water. The research investigated perceptions of four water supply options: current practice (treated wastewater is discharged into rivers and becomes drinking water for downstream communities); blended reservoir; upstream discharge (treated wastewater is discharged into the river upstream from the community, and hence drawn upon for reuse); and direct potable reuse.

The authors found that blended reservoir was preferred, followed by upstream discharge, direct potable, and current practice.

More recently, Fielding, Gardner, Leviston, and Price (2015) compared Australians' comfort with drinking four alternative water sources across four studies. They found that comfort with drinking rainwater was highest, and recycled water lowest, with desalinated water and stormwater in between. They did not provide participants with a definition of each water source. Factors found to be significant positive predictors of comfort with drinking the alternative water sources investigated included participants' trust in science and the government, and their comfort with technology in general. Leonard, Mankad, and Alexander (2015) compared community attitudes to three alternative water supply systems using stormwater through managed aquifer recharge. These were a third-pipe (nonpotable) system; an indirect potable system, where the water is recovered from the aquifer, transported to the drinking water reservoir and treated with the existing water supply; and direct potable use, where the stormwater is recovered from the aquifer to a local treatment plant and then distributed in the drinking water mains. In total, 1043 respondents representative of the greater Adelaide population were surveyed; 73% supported the nonpotable use, 57% the indirect potable use, and 55% the direct potable use.

Overall, only a limited number of studies have compared public attitudes to more than one alternative water source at the one time period. Furthering this work would be beneficial.

Perceptions of alternative water sources

There is a small but growing body of literature exploring public perceptions of the attributes (e.g. aesthetics, and health and safety) of different water sources, including tapwater (Bruvold, 1968; Doria, 2010) and recycled water (Hurlimann & McKay, 2007). More recently, studies in Australia have compared public perceptions of multiple sources of water and found significant differences between sources (Dolnicar & Schäfer, 2009; Dolnicar et al., 2014). For example, Dolnicar et al. (2014) investigated Australian perceptions of five water sources: bottled water; current tap water; rainwater from a tank; desalinated water; and recycled wastewater. They found a significant difference in the evaluations respondents made between each water source for a wide range of desirable and undesirable water attributes. Bottled water and tapwater were perceived as healthy and safe for human consumption. Recycled water was perceived as the worst-performing source of water in terms of aesthetic attributes and health concerns. To the present authors' knowledge there has not been a study to assess public perceptions of a range of water sources and compare these across multiple locations.

Location differences in public acceptance

Only a limited number of studies have compared public acceptance of alternative water sources across locations. Roseth (2008) compared attitudes to recycled water across purposes in five Australian locations (Adelaide, Brisbane, Melbourne, Perth and Sydney). Brisbane respondents were more willing to use recycled water for a range of purposes. Respondents from Adelaide were less willing to accept the use of recycled water in the cooking industry, for household pools, for cooking and for drinking. Macpherson and Snyder's (2013) study, discussed earlier, also involved the quantitative comparison of attitudes to the alternative water supply options across two survey samples – one in the US and the other in Australia.

They found differences in preferred water source between respondents from those two locations. The report indicates that the majority of respondents in both countries were "willing to" drink or "generally OK" with each of the four scenarios presented. More Australians were "willing to" drink the water from each scenario. However, no information was provided on whether the difference was statistically significant.

The present study contributes to this body of work by comparing three alternative water sources for seven purposes across nine locations, in nine countries.

Method

An international online survey was conducted in June 2012 in nine countries: Australia, Belgium, Canada, Israel, Japan, Jordan, Mexico, Norway and the US (specifically Los Angeles, California). The purpose was to explore how acceptance and perceptions of recycled water, desalinated sea water and rainwater varied between these nine diverse locations and their associated diverse water, cultural and social contexts.

Locations

The locations were chosen to be diverse with respect to water scarcity, water supply, political system, general societal environmental approach and outlook, and climatic and socio-economic factors. Values for each location included in the sample are presented in Table 1 – data at the national level. It should be noted that these indicators were used as a guide only, given that some of the sources used to compile Table 1 are dated. For example, the FAO (2013) figures regarding water supply source for each location are from the early 2000s; desalinated and recycled water production capacity has increased in many countries since then.

Long-term average annual precipitation was taken from the FAO's online database, AQUASTAT (FAO 2013).

Total water withdrawn was taken from AQUASTAT.

Pressure on water resources was taken from the AQUASTAT. The database calculates pressure on water resources by establishing the total freshwater withdrawn by each nation and expressing it as a percentage of the nation's annual renewable water resources.

Water supply. Given the limitations on the *pressure on water resources* measure, the broader water supply conditions of the locations were also considered, including historical water supply shortfalls and water supply source, to ensure a diversity of water supply conditions in the sample (see the Water Supply column in Table 1). An additional column details the water supply source for each location, also from AQUASTAT.

Population density was taken from AQUASTAT and is measured in inhabitants per km².

Political system was drawn from the Freedom House (2011) Freedom in the World index, which is based on two measures: political rights and civil liberties. Each of these is measured on a scale ranging from 1 (most free) to 7 (least free). Countries are then grouped into three categories: free (less than 2 on each measure); partially free (3–5 on each measure); and not free (more than 6, or mixed scores of one 5 and one 6 or 7).

General societal environmental approach and outlook was measured with the *social and institutional capacity* component of the ESI 2005 (Yale Center for Environmental Law & Policy, Yale University Center for International Earth Science Information Network, Columbia University,

Table 1. Sample country characteristics.

Country/Water supply	Population density[1] (inh/km²)	Political System Score[2]	Environmental approach[3]	Average daily min./max. temperature (°C), 1961–90[4]	Long-term average annual precipitation[1] (mm/y)	Total water withdrawal[1] (m³ per capita per year)	Pressure on water resources[1,6]	Life expectancy at birth, 2010[5]	PPP GNI/cap[5]	Access to improved water supply (% of pop.)[5]	Water supply profile[1,7]
Australia Many capital cities facing challenges to water supply. Significant drought in the 2000s in many regions of the country. Augmentation of supply with multiple sources being implemented or considered in numerous locations (Grant et al., 2013).	3	Free	17	14.9/28.4	534	1,152	4	82	36,910	100	S = 73%; G = 25%; R = 2%; D ≤ 1%
Belgium Potable reuse through aquifer recharge, e.g. in Torreele (Water Supply and Sanitation Technology Platform, 2010)	352	Free	20	5.5/13.6	847	589	34	80	38,240	100	S = 90%; G = 10%; R ≤ 1%
Canada Historically, plentiful supply of fresh water and high per capita use. Growing awareness of need for sustainable water management due to increasing water use, population and pollution (Environment Canada, 2011; Fennell and Kielbasinski, 2014).	3	Free	16	−10.1/−0.6	537	1,589	1	81	38,370	100	S = 96%; G = 4%; D ≤ 1%
Israel Low per capita freshwater availability. Diverse range of projects to address this water scarcity, including adoption of water recycling for agricultural reuse, and more recently desalination of seawater for potable use (Alon, 2006).	342	Free	30	13.4/25.0	435	282	79	82	27,660	100	R = 13%; D = 7%; S&G = 80%

(continued)

Table 1. (*continued*).

Country/Water supply	Population density[1] (inh/km²)	Political System Score[2]	Environmental approach[3]	Average daily min./max. temperature (°C), 1961–90[4]	Long-term average annual precipitation[1] (mm/y)	Total water withdrawal[1] (m³ per capita per year)	Pressure on water resources[1,6]	Life expectancy at birth, 2010[5]	PPP GNI/cap[5]	Access to improved water supply (% of pop.)[5]	Water supply profile[1,7]
Japan Non-potable reuse occurs (Kimura et al., 2007; Yamagata et al., 2002). Frequent water shortages in many areas, with reduced water pressure and hours of supply a measure during these periods (Morimasa et al., 2014).	335	Free	5	7.0/15.3	1,668	714	21	83	34,610	100	S = 82%; G = 18%; D ≤ 1%; R ≤ 1%
Jordan Arid/semi-arid. High water stress in terms of availability of water per capita, and population growth. Unreliable water supply based on rotation and prone to failure. Alternative water sources sought by individuals, including rainwater (Abdulla & Al-Shareef, 2009).	71	Not free	52	11.2/25.4	111	166	99	73	5,800	97	R = 9%; D = 1%; S = 31%; G = 59%
Mexico Mexico faces a challenging water future (Spring, 2011). Mexico City's water supply is significantly stressed; extraction of water exceeds natural availability, yet consumption is amongst the highest in the nation (Novelo & Tapia, 2011).	58	Partly free	59	13.5/28.5	752	700	17	77	14,340	96	R = 2.5%; S = 60%; G = 37.5%

Norway	Plentiful supply of freshwater. Hydroelectric power generation and flooding are key future concerns (Lawrence & Haddeland, 2011; Norwegian Ministry of Petroleum and Energy, 2015).	15	Free	3	−2.0/5.0	1,414	622	1	81	58,570	100	S&G = 100%
United States	Many areas face increasing water scarcity; consumption per capita is high; potable reuse being considered and implemented in numerous areas (National Research Council, 2012). Potable reuse through aquifer recharge in parts of the country. The focus of the US sample is on Los Angeles, given its future water challenges (Wetterau et al., 2011)	32	Free	14	2.2/14.9	715	1,583	16	78	47,310	99	S = 77%; G = 23%; R ≤ 1%; D ≤ 1%

[1] Food and Agriculture Organization of the United Nations (2013).

[2] Freedom House (2011) and above discussion.

[3] Yale Center for Environmental Law and Policy, Yale University Center for International Earth Science Information Network, Columbia University, World Economic Forum, Joint Research Centre and European Commission (2005); and above discussion.

[4] OECD (2008).

[5] Purchasing power parity gross national income per capita, 2010 dollars, from World Bank (2012).

[6] Total water withdrawn as a percentage of annual renewable water resources (Food and Agriculture Organisation of the United Nations, 2013).

[7] R = treated wastewater; D = desalinated water; S = surface water; G = groundwater.

World Economic Forum, Joint Research Centre, & European Commission, 2005). This comprises four indicators: *science and technology*; *environmental governance*; *eco-efficiency*; and *private-sector responsiveness*, each containing 2–12 variables. The 146 countries included in the ESI are then ranked from 1 to 146, where 1 is the best.

Average daily minimum and maximum temperature was sourced from the 2006–2008 OECD Environmental Data Compendium (OECD, 2008).

Life expectancy at birth was gathered from the World Bank Development Indicators (World Bank, 2012), using the most recent data available at the time of publication (2010).

Income indicators were gathered from the World Bank Development Indicators (World Bank, 2012, p. 23) using the measure of *purchasing power parity gross national income per capita*. For this measure, the data are converted to international dollars using purchasing power parity rates, and divided across the population (calculated mid-year).

Study locations were selected so as to ensure that a diverse range of water-related conditions were represented. Also, the online fieldwork company had to be able to provide a sufficiently large representative sample for each of the countries ultimately included.

Participant recruitment

A research-only, permission-based Internet panel recruited 200 participants in each of the nine locations. The sample size per location was relatively small because a compromise was necessary between cost and margin of error. It was determined that a 7% margin of error for the worst-case scenario of a 50% estimated population proportion was acceptable (Cochran, 1977). Data were collected in June and July 2012. The Melbourne office of an international marketing research consultancy engaged partners they had in the other eight locations. The consultancy was instructed to recruit a nationally representative sample (age, gender, location), except in Los Angeles, where it was to be representative of Los Angeles. They were asked to draw a sample of respondents from neighbourhood/regional socio-economic quotas in each location.

The survey was made available in the official language of each location, with professional translators used to translate the survey from its original English version. The survey took approximately 30 minutes to complete. Respondents were paid a small compensation fee for their efforts.

Survey

Respondents were asked a series of questions about their attitudes to, perceptions of, and behaviours relating to water. They were asked additional questions about their general environmental attitudes and behaviours, living conditions, and socio-economic information. In designing the survey, the researchers were mindful to use concepts that could be understood across nations. For this reason, early in the survey, respondents were provided with a short statement of information about the various water sources discussed in the survey, to ensure a common understanding across respondents (Box 1). The information also sought to connect the water sources with locations already using that source, to ground the scenario in a real-world example. This was done to address one limitation of this study, its hypothetical nature.

Box 1: Information about water sources provided to survey participants

Throughout the survey:

- We will use the term **"rainwater"** to describe **"rainwater from a rainwater collection tank on your property"** (rainwater collected from the roof of your house)
- We will use the term **"bottled water"** to describe "**water sold in bottles by food companies that is widely available to the public for purchase and consumption"**
- We will use the term **"your current tap water supply"** to describe the **"water you presently use throughout your dwelling (home)"**
- We will use the term "**recycled water**" to describe "**highly purified wastewater deemed by scientists as safe for human consumption"**. Such water is currently used for drinking purposes in countries including the USA, Singapore and Namibia.
- We will use the term **"desalinated water"** to describe **"highly purified seawater deemed by scientists and public health officials as safe for human consumption."** Such water is currently used for drinking purposes in countries including the USA, Australia and the Middle East.
- And we will assume that both recycled and desalinated water were treated to the same level of water quality.

The survey investigated stated willingness to use three alternative water sources – recycled water, desalinated water and rainwater – for seven purposes. Respondents were asked:

> For the following question, imagine that you live in a town that is facing a water shortage. Please indicate whether or not you would use <water source> for the following purposes.

Respondents were presented with a list of seven water use purposes (watering plants, washing clothes, washing my body, drinking, toilet flushing, cooking, and cleaning) for each water source. Response options were yes or no. Binary options were offered deliberately, given the cross-cultural nature of the study. It is well known that respondents from different cultural background use response options differently. These differences in "response styles" (Paulhus, 1991) manifest in the data-set as biases which can lead to misinterpretation of findings. Binary answer options eliminate the possibility of most biases occurring (Rossiter, Dolnicar, & Grün, 2015).

Respondents were also asked about their perception of five water sources (purified wastewater, purified seawater, tapwater, bottled water and rainwater):

> You will now see a list of descriptions of water. For each kind of water, please indicate whether or not they apply. If you are not sure, please tick the option you think is more likely.

Respondents were then presented with a list of nine perception statements (presented in random order for each respondent, except for the 'knowledge' question, which was fixed at last). The perception statements were originally used by Dolnicar et al. (2014) in an Australian study, and related to aesthetic perceptions, cost, environmental impact, convenience and knowledge. The response options were yes or no.

Respondents were also asked what water source they use for drinking. They were presented with six water sources – town-supplied water, rainwater tank, bottled water, groundwater, recycled water and desalinated water – in addition to the options "I boil my water", "other", and "don't know". Multiple responses were allowed, except when "don't know" was chosen. Boiling water was included in this research because it represents a common practice to improve water quality (Katuwal & Bohara, 2011; Sodha et al., 2011). Additionally, it is

Table 2. Drinking water source – percentage of respondents who indicated they used each water source for drinking purposes in each location.

Country/Region	Town supplied water %	Rainwater from tank %	Bottled water %	Ground water %	I boil my water %	Recycled water %	Desalinated water %	Other %	Don't know %
Belgium	55	5	64	3	1	2	0	3	0
Norway	85	1	19	8	4	2	1	2	4
Israel	56	1	47	3	8	0	3	15	2
Australia	75	10	27	1	10	0	1	7	3
LA (USA)	45	1	61	3	8	2	1	8	3
Canada	68	1	45	8	8	3	1	8	2
Mexico	56	13	37	23	4	7	4	3	4
Japan	69	1	46	7	10	1	0	9	2
Jordan	30	19	67	11	14	17	7	4	1

▨ most stated water source. ▢ second-most stated source. ☐ third-most stated source.

common practice for boil-water alerts to be issued when centralized water supply quality has been compromised (Hrudey & Hrudey, 2006).

Results and discussion

Sources of drinking water currently utilized

Sources of water used by respondents for drinking are shown in Table 2. Dark-grey shading indicates which water source is used by the largest percentage of respondents in each location for drinking; medium grey indicates the second-largest source of drinking water, and light grey the third-largest. Town-supplied water was the dominant source of drinking water for respondents in six of the countries surveyed. Bottled water was most frequently the main drinking water source in Jordan, Belgium and Los Angeles. Only a few respondents indicated they drink recycled or desalinated water. The source of their current centralized water source was not captured in the survey.

Jordan had the most diverse drinking water sources stated by respondents. In addition to bottled water (67%), 30% of Jordanian respondents indicated they drank town-supplied water, 19% water from a self-supplied rainwater tank, 17% recycled water, and 11% groundwater. This diversity of water sources may be due to the unreliability of town-supplied water, and gaps in supply, as reported by Abdulla and Al-Shareel (2009). The diversification of drinking water supplies is in line with that employed by respondents in Özdemir et al.'s (2011) study in the Mekong Delta. Additionally, Jordan had the highest percentage of respondents who reported boiling drinking water (14%), above Australia and Japan (10% each), which again may be reflective of the real or perceived issues surrounding water quality in many locations in the country, as boiling water (and doing so properly, with proper storage) is an important public health measure (Rosa & Clasen, 2010). While town-supplied water was the main source of drinking water indicated by the respondents from Mexico, 37% also drank bottled water, 23% groundwater, and 13% rainwater from their own tank. As in Jordan, this may be due to lower reliability of centralized supply in Mexico compared to other countries. As indicated in Table 1, full access to 'improved water sources' for Mexico's and Jordan's populations has not yet been reached.

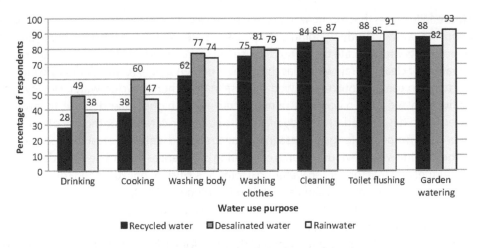

Figure 1. Intended use of recycled water, desalinated water and rainwater for various purposes – whole sample.

In Australia, 75% of respondents indicated they drank town-supplied water, the second-highest percentage, behind Norway (85%). However, 27% of Australians indicated they drink bottled water, and 10% rainwater. A previous Australian study found that 42% of people In South Australia drink rainwater due to concerns about aesthetic and chemical properties of tapwater (Heyworth, Maynard, & Cunliffe, 1998). Norway had the lowest number of respondents indicating they drink bottled water (19%), while 8% indicated they drink groundwater.

Overall, the results indicate that a diversity of water sources is drawn upon for drinking in each of the countries surveyed, with tapwater and bottled water the dominant sources overall. These results may be reflective of the rise in bottled water sales globally over the past two decades (Parag & Roberts, 2009). This study did not assess frequency of use of the stated drinking water sources, so the conclusions drawn here are limited in this regard. This means it is not known whether the 27% of Australian respondents drink bottled water seldom or regularly.

Intended use of alternative water sources

Intended use of recycled water, desalinated water, and rainwater from a tank were first analyzed at the whole-sample level; then differences between locations were explored. The percentage of respondents (whole sample) who indicated they would use each alternative water source for the seven purposes is shown in Figure 1. The overall percentage of respondents who indicate they intend to use recycled water and rainwater increases with decreasing physical contact with the water. For recycled water, this is in line with findings from previous research (e.g. Alhumoud & Madzikanda, 2010; Browning-Aiken et al., 2011; Bruvold, 1988; Marks et al., 2006; McKay & Hurlimann, 2003; Sydney Water, 1996). Overall, this trend holds for desalinated water, but it receives a lower preference for watering the garden than for toilet flushing or cleaning. Toilet flushing receives the same rating as cleaning. Statistical tests (chi square) were undertaken to establish whether there was a significant difference between water sources for each water use purpose. All were significant.

Figure 1 illustrates that for the water uses which have close-to-person applications (drinking, cooking, washing body, washing clothes), desalinated water had the highest percentage of respondents who indicate they intend to use the source, followed by rainwater, then recycled water. For cleaning, rainwater has the highest percentage of respondents who indicate they intend to use the source, followed by desalinated, then recycled water. However, for toilet flushing and garden watering, rainwater received a higher intended use, followed by recycled, then desalinated. This study places the stated preference for the use of rainwater for a range of purposes alongside desalinated and recycled water. The results reported here are largely in line with previous research comparing preferences for recycled and desalinated water, suggesting that for garden watering (Dolnicar & Hurlimann, 2010) and toilet flushing, recycled water is the preferred source (Dolnicar & Hurlimann, 2010; Dolnicar & Schäfer, 2009).

Table 3 presents results by location. The table uses shading to indicate which water source (recycled, desalinated or rainwater) is most accepted (dark grey), second-most accepted (medium grey), and third-most accepted (light grey) for each location. The chi-square test was used to establish whether the observed differences across locations is random, for each water source / use application. Given the large number of independent tests being run (9 locations × 7 water sources = 63 tests), there is a chance that the significance of each item is overestimated; thus, the p-values were Bonferroni-corrected to account for multiple testing. The results indicated that the observed differences for each water source/use tested were not random but significant at the .01 level, except for three uses: desalinated water for body washing; rainwater for cleaning; and rainwater for watering plants.

Water perceptions across five water sources

Table 4 shows, by location, the percentage of respondents who agreed with each perception statement, for the five water sources included in this question. The chi-square test (Bonferroni-corrected) was used to establish whether the observed differences across locations is random for each water source / perception statement. The results indicate that the observed differences for each water source / perception statement tested were not random but significant at the .01 or .05 level, except for 'is expensive' for 'rainwater from a tank', which was not significant.

While there were differences between locations in the percentage of respondents who agreed with each statement, there were some similarities in terms of water source perception order. For example, in comparison to the other water sources in the survey, bottled water was perceived as the cleanest water source by the highest proportion of respondents in every location. Bottled water was also perceived as the least harmful to people's health of all water sources in all locations except Norway, where it was perceived as the second-least harmful overall. Bottled water was perceived as the most expensive source of water in all locations, and the least environmentally friendly in all locations but Mexico and Jordan.

Additionally, rainwater was perceived as the least expensive source of water in all locations. It was also perceived as the most environmentally responsible source of water in all but two locations (Norway and Mexico), and the least clean source of water in all but two locations (Israel and Jordan). Current water supply was rated the most convenient source of water in all but three locations – Mexico, Japan and Jordan. For recycled water and desalinated water, respondents had the lowest stated knowledge of these water sources, and they were seen as the least convenient. While the percentage of respondents agreeing

Table 3. Percentage of respondents reporting intended use of recycled water (R), desalinated water (D), and rainwater from tank (T), in a water-shortage scenario, by location.

Water Use	Watering plants % yes			Washing clothes % yes			Washing my body % yes			Drinking % yes			Toilet flushing % yes			Cooking % yes			Cleaning % yes		
Water Type / Location	R*	D*	T	R*	D*	T	R*	D	T*	R*	D*	T*	R*	D*	T*	R*	D*	T*	R*	D*	T
Belgium	86	81	96	79	81	80	67	74	63	28	34	16	86	84	96	41	53	24	86	85	92
Norway	87	81	91	84	84	88	75	78	86	32	46	39	88	86	92	48	60	56	83	86	88
Israel	95	94	94	76	88	82	59	81	76	13	52	45	92	90	91	31	65	53	88	92	87
Australia	96	84	91	75	87	86	56	80	83	33	59	66	94	92	92	42	68	73	86	89	87
LA (USA)	93	85	96	74	83	82	53	74	69	26	40	34	95	91	93	39	57	44	86	92	87
Canada	89	80	95	73	75	80	58	74	70	25	48	26	90	83	91	35	58	44	83	84	83
Mexico	86	79	95	79	82	86	67	82	82	32	62	43	88	83	95	42	67	51	86	84	95
Japan	85	75	94	73	76	52	60	72	47	24	42	10	85	85	91	25	45	11	85	83	84
Jordan	77	80	87	61	74	81	60	79	88	37	59	61	73	67	79	40	70	67	68	76	83
Average	88	82	93	75	81	79	62	77	74	28	49	38	88	85	91	38	60	47	84	85	87

*Uses and water types significantly different between countries at the .01 level using chi-square test.
For each country the acceptance responses for each water use purpose have been shaded as follows: ■ most accepted source; ▨ second-most accepted source; ☐ third-most accepted source.

Table 4. Percentage of respondents reporting perceptions of recycled water (R), desalinated water (D), rainwater from tank (T), current water supply (C), and bottled water (B), by location.

Perception about water source / Water Type / Location	Is environmentally friendly (% yes)					Is potentially harmful to people's health (% yes)					Is expensive (% yes)					Is clean (% yes)					I know a lot about this kind of water (% yes)					Is convenient (% yes)				
	R*	D*	T*	C*	B*	R*	D*	T*	C*	B*	R*	D*	T	C*	B*	R*	D*	T*	C*	B**	R*	D*	T*	C*	B*	R*	D*	T*	C*	B*
Belgium	79	77	85	81	63	42	29	52	18	8	58	60	4	56	81	66	70	30	84	96	11	12	43	48	54	67	58	85	93	88
Norway	69	63	80	91	41	51	31	54	13	14	52	55	7	14	87	46	62	34	84	86	14	14	30	62	54	41	47	58	88	74
Israel	79	82	94	82	44	66	28	24	33	21	50	68	8	45	91	36	72	68	61	88	10	18	37	56	53	41	54	51	94	66
Australia	70	56	96	72	30	44	27	31	33	22	75	80	8	14	88	66	76	68	75	91	20	17	67	62	50	35	32	71	97	76
LA (USA)	75	77	85	73	60	53	37	43	45	18	64	75	7	24	76	55	63	47	58	89	17	16	45	63	69	33	27	43	93	86
Canada	76	70	85	67	44	52	34	55	32	24	53	68	10	16	77	57	66	38	79	90	15	11	47	72	65	38	38	62	93	84
Mexico	35	28	21	29	32	53	32	36	54	10	38	66	9	36	81	55	66	47	51	96	18	15	54	54	68	58	61	69	66	89
Japan	58	67	74	67	54	75	42	80	40	20	40	62	5	24	89	25	49	7	81	92	4	6	12	29	28	20	23	35	91	96
Jordan	45	73	97	68	77	81	43	26	64	33	54	54	12	33	93	24	58	76	47	88	38	38	81	69	81	25	51	83	57	93
Average	65	66	80	70	49	57	34	45	37	19	54	65	8	29	85	48	65	46	69	91	16	16	46	57	58	40	43	62	86	84

*Uses and water types significantly different between countries at the .01 level using chi-square tests; ** .05 level.

The country with the highest rate of agreement per water source for each perception statement is indicated in **bold**.

The country with the lowest rate of agreement per water source for each perception statement is indicated with underline.

For each country the acceptance responses for each water use purpose have been shaded as follows: ■ highest agreement with perception statement; □ second-highest; ▨ third-highest.

with each perception statement differed between locations, the average agreement scores across locations indicates that recycled water was perceived by the highest proportion of respondents as potentially the most harmful to human health, and the second-least clean source of water – behind rainwater from a tank.

In order to explore whether there is a correlation between perceptions of the water sources and willingness to use them, Pearson correlation analysis was undertaken. Given the large number of perception statements investigated in this study, and the number of alternative water sources, the analysis was undertaken separately for each kind of water for only one water use (drinking) and one perception statement: "<alternative water source> is potentially harmful to human health". The analysis was undertaken for the overall sample. The perception statement for each water source was tested against willingness to use this source for drinking. A significant ($p = .01$) negative correlation was found between the perception that the alternative water source is potentially harmful to human health and the willingness to use it for drinking (recycled water $r(1791) = -0.169$; desalinated water $r(1791) = -0.21$; rainwater $r(1791) = -0.279$). A role is indicated for water authorities to improve communication about potential risks to human health associated with various water sources, to allay community concerns. Further research to explore this would be beneficial.

Discussion of stated water preferences and perceptions in specific locations

Australia

The water source Australians preferred depended on purpose (Table 3). Desalinated water was preferred for drinking and cooking. In comparison to other locations, Australians were most willing to use desalinated water for cooking. This may be due to the fact that desalinated water was seen by Australian respondents as the cleanest and least harmful to human health, behind bottled water. This was the location most willing to use recycled water on the garden, and the most willing to drink and cook with rainwater, perhaps reflective of a cultural history of drinking from rainwater tanks (Heyworth et al., 1998) and a high penetration of rainwater tanks in some contexts (e.g. 41% in Victoria, Hurlimann, 2011). Additionally, the results in Table 4 indicate Australians perceive rainwater as the most environmentally friendly source of water, and it is the water source for which they state they have the highest level of knowledge.

Belgium

Belgian participants preferred desalinated water for the four most personal uses investigated, and rainwater for the three least personal (Table 3), with the latter also the water source which the largest percentage of respondents perceived as environmentally friendly. In comparison to other locations, Belgian study participants were the least willing to use desalinated water for drinking purposes and the most willing to use rainwater for toilet flushing. This may be explained by the fact that this was the location with the largest percentage of respondents perceiving rainwater from a tank as convenient to use.

Canada

Canadians preferred desalinated water for all purposes except washing clothes and toilet flushing, where rainwater was preferred (Table 3). Compared to rainwater and recycled wastewater, a smaller percentage of respondents viewed it as potentially harmful to human health, and a higher percentage viewed it as clean. Rainwater was perceived as the most environmentally friendly source (Table 4).

Israel

Overall, desalinated water was the preferred source of water for Israelis for all purposes except watering plants and toilet flushing, where recycled water was the preferred choice (Table 3). In comparison to other locations, Israelis preferred recycled water the most for cleaning, but the least for drinking; they viewed recycled water as the most risky to human health. In comparison to other locations, Israelis preferred desalinated water the most for washing clothes, cleaning, and watering plants, and this was the location with the highest percentage of respondents viewing desalinated water as environmentally friendly (Table 4).

Japan

For Japanese respondents, the preferred source of water varied across the purposes of use (Table 3). In comparison to other locations, they were the least willing to use recycled water for cleaning, and the least willing to use desalinated water for cooking, washing the body and watering plants. A very low percentage of Japanese respondents perceived these water sources as clean or convenient to use. Additionally, in comparison to other locations, they were the least willing to use rainwater for the four most close-to-person uses, which could be attributed to the Fukushima nuclear incident which occurred in March 2011. They had the lowest stated knowledge levels of all five water sources.

Jordan

For Jordanians, rainwater was the preferred water source for all uses except drinking and cooking, where desalinated water was preferred (Table 3). In comparison to other locations, they were the least willing to use rainwater for toilet flushing or watering plants. As seen in Table 4, Jordanians perceived rainwater as the cleanest water source, and the least harmful to people's health. Additionally, in comparison to other locations, Jordanians were the most willing to use recycled water for drinking. This was the location least willing to use desalinated water for washing clothes, cleaning, and toilet flushing, but the most willing to use it for cooking. In comparison to other locations, they had the highest knowledge levels for all water sources except for current tapwater. The overall pattern of willingness to use the three water sources across the seven water use purposes which was observed for Jordan appears to indicate a frugal attitude to water use, reflective of their water-stressed context (see Table 1). The pattern of increasing acceptance of recycled water use observed for the whole sample holds for Jordan, but it does not hold for desalinated water or rainwater. Overall, willingness to use each alternative source decreases for low-human-contact uses. In comparison to other locations, their current tapwater is viewed by the lowest percentage of respondents as clean, or convenient, and the perception that it is potentially harmful to human health is the highest of all locations. Further exploration, particularly of a qualitative nature, of Jordanian attitudes to water in this respect would be valuable.

Mexico

For the Mexican respondents, rainwater was the preferred source for all uses except drinking and cooking, where desalinated water was the preferred choice, in addition to washing the body, where it was equal with rainwater (Table 3). In comparison to other locations, they were the most willing to use desalinated water for drinking and for body washing, and were the most willing to use rainwater for cleaning. As detailed in Table 4, this was the location with the highest proportion of respondents perceiving desalinated water as convenient, and the second-lowest percentage viewing their current tapwater as convenient, behind Jordan. They had the lowest percentage of respondents viewing each water source as environmentally friendly.

Norway

Norwegians preferred rainwater for all uses except for drinking and cooking, for which desalinated water was preferred. In comparison to other locations, Norwegians were the most willing to use recycled water for cooking, washing the body and washing clothes, despite being the least water-stressed location in the study (Table 1). This may be due to their high score in terms of environmental approach (Table 1). In comparison to the other locations, they were the most willing to use rainwater for washing clothes. This was the location with the most positive perceptions of the current water supply (Table 4), with a high percentage of respondents viewing it as clean and environmentally friendly, and the lowest percentage viewing it as potentially harmful to human health, or expensive.

Los Angeles (California, USA)

For the Los Angeles sample, desalinated water was the preferred water source for all uses except for watering plants, where rainwater was preferred, and toilet flushing, where recycled water was preferred. This may be due to the fact that desalinated water was perceived by the Los Angeles sample as the second-safest source of water (for human health), behind bottled water. In comparison to other locations, they were the most willing to use recycled water for toilet flushing, but the least willing to use it for washing the body.

Conclusions

There is an increasing acknowledgement that traditional centralized water supply systems need to adapt to existing and future challenges, including climate change. One approach to this challenge is to diversify the water supply to include the use of alternative, nontraditional sources such as desalinated seawater, recycled wastewater and the more widespread use of rainwater tanks in developed nations. Yet, there is no guarantee that these changes will be accepted by the public involved. Thus, further research is needed to understand public opinion regarding these alternative water sources. The study reported in this article addresses this research need.

Across the total sample of 1800 participants, the study found the ordered preference for recycled water and desalinated water across a range of water use purposes to be consistent with previous studies (Dolnicar & Hurlimann, 2010; Dolnicar & Schäfer, 2009). Desalinated water was preferred for all uses except the two least personal ones (garden watering and toilet flushing). Importantly, the study included a comparison with rainwater and found that rainwater was the preferred water source (over recycled water and desalinated water) for the least personal water uses: cleaning, toilet flushing and garden watering. For all other uses rainwater was the second preferred source, behind desalinated water.

The results indicate that survey respondents differentiated between the various water sources and their applied use. Desalinated water, a purified and scientifically treated source (as defined in Box 1), was preferred over rainwater for close-to-person uses. But rainwater was preferred to recycled water that was "purified and scientifically tested" – despite the increased health risks in untreated rainwater consumption. It is possible that participants considered the option of a point-of-consumption treatment intervention such as boiling water, which is common practice in many locations across the globe (Rosa & Clasen, 2010). Further research to explore the reasons behind these preferences would be of interest, particularly in light of Hurliman and Dolnicar's (2010) research in Toowoomba, Australia, which indicated some residents were of the view that if recycled water were provided they would simply substitute another source for drinking.

The study offers insights into differences in stated willingness to use alternative water sources, and differences in water source perceptions, across nine locations. Importantly, significant differences emerged for preferred water source for all but three of the water source/ purpose of use combinations researched. While the percentage of respondents willing to use each water source for each purpose varied between nations, what was consistent across most locations was a higher willingness to use alternative water sources for less personal uses of water. The exception is Jordan, where the aggregated results show a more nuanced approach to considering the use of alternative water sources, perhaps reflective of the precarious water situation there, and exposure to a more diverse array of water sources.

This research into perceptions of five water sources across nine locations found differences between locations in the percentage of respondents who agreed with each perception statement for most water sources. However, consistent across locations was the perception that bottled water is the cleanest of the water sources investigated. Also consistent was the order of perceived expense of water source: bottled water, followed by desalinated water, recycled wastewater, current water supply, and finally rainwater from a tank. Importantly, this study finds that knowledge about recycled water and desalinated water was low across all locations in the study, suggesting the value of greater engagement with the community about these water sources. The results suggest that effective communication with the public regarding potential public harm will add to willingness to use alternative water sources – given the negative correlation found between such perceptions and willingness to use each alternative water source.

Regional variations may be expected in attitudes to alternative water sources within the countries and locations sampled. This is because water context varies greatly within some of the locations included in this study. For example, results for the general population of the US would be very different from the results for Los Angeles, given this city's specific water and cultural context.

The research results reported in this article provide insights which may be of value to water supply policy makers, and specifically those companies working in diverse locations and across national boundaries. The results of this study show that attitudes and perceptions of alternative water sources vary across the locations studied. Importantly for water policy makers and multinational water companies, the results indicate that investing in social research with the communities at the heart of water supply problems and decisions will be important – it cannot be assumed that the results of a study done in one location will hold in another location. The results of the attitude and perception statements found in Table 5 can serve as a starting point for water communication strategies.

There are three key limitations to this study. First, the sample size of 200 study participants per location is relatively modest. As a consequence, the population percentages presented, as in any such study, may lie slightly above or below the real percentages, by the margin of error (±7%). Second, while every effort was made for samples to be representative, it has to be acknowledged that this study was conducted in very diverse locations, which have different census statistics with different levels of reliability. Some error may have resulted from such differences. More extensive research in each location would be beneficial. Finally, this study uses stated behaviours and stated behavioural intentions. These are not always the same as actual behaviour. Actual behaviour, however, is impossible to measure when such a large number of locations is compared and not all alternative water sources are available in all locations. Nevertheless, this represents a key limitation which needs to be taken into consideration when interpreting the findings.

Acknowledgements

Research assistance contributing to this article was provided by Mathew Caulkins, Hannah Kelly and Yogita Rijal. Thanks to Lindsay Tanner, librarian at Oxford Brookes University, for assistance in finding sources for data in Table 1 while Anna Hurlimann was a visiting scholar at Oxford Brookes University on sabbatical in 2013.

Disclosure statement

No potential conflict of interest was reported by the authors.

Funding

This research was funded by Australian Research Council (ARC) grant DP0878338. Salary funding for Dolnicar was provided through ARC grant DP110101347.

References

Alon, T. (2006). Seeking sustainability: Israel's evolving water management strategy. *Science, 313*(5790), 1081–1084.

Abdulla, F. A. & Al-Shareel, A. W. (2009). Roof rainwater harvesting systems for household water supply in Jordan. *Desalination, 243*, 195–207.

Aitken, V., Bell, S., Hills, S., & Rees, L. (2014). Public acceptability of indirect potable water reuse in the south-east of England. *Water Science & Technology: Water Supply, 14*, 875–885.

Alhumoud, J. M. & Madzikanda, D. (2010). Public perceptions on water reuse options: The case of Sulaibiya wastewater treatment plant in Kuwait. *The International Business & Economics Research Journal, 9*, 141–158.

Al-Mashaqbeh, O. A., Ghrair, A. M., & Megdal, S. B. (2012). Grey water reuse for agricultural purposes in the Jordan Valley: Household survey results in Deir Alla. *Water, 4*(4), 580–596.

Barthwal, S., Chandola-Barthwal, S., Goyal, H., Nirmani, B., & Awasthi, B. (2014). Socio-economic acceptance of rooftop rainwater harvesting – A case study. *Urban Water Journal, 11*, 231–239.

Biswas, A. K. (2008). Integrated water resources management: is it working? *International Journal of Water Resources Development, 24*, 5–22.

Browning-Aiken, A., Ormerod, K. J., & Scott, C. A. (2011). Testing the climate for non-potable water reuse: opportunities and challenges in water-scarce urban growth corridors. *Journal of Environmental Policy & Planning, 13*, 253–275.

Bruvold, W. H. (1968). Scales for rating the taste of water. *Journal of Applied Psychology, 52*, 245–253.

Bruvold, W. H. (1972). *Public attitudes toward reuse of reclaimed water.* California: University of California, Water Resources Centre.

Bruvold, W. H. (1988). Public opinion on water reuse options. *Journal of the Water Pollution Control Federation, 60*, 45–49.

Bruvold, W. H. (1992). Public evaluation of municipal water reuse alternatives. *Water Science and Technology, 26*, 1537–1543.

Bruvold, W. H. & Ward, P. C. (1970). Public attitudes toward uses of reclaimed wastewater. *Water and Sewage Works, 117*, 120–122.

Carr, G., Potter, R. B., & Nortcliff, S. (2011). Water reuse for irrigation in Jordan: Perceptions of water quality among farmers. *Agricultural Water Management, 98*, 847–854.

Cochran, W. G. (1977). *Sampling techniques* (3rd ed.). New York, NY: John Wiley & Sons.

Davies, A. (2006, February 8). Desalination plant dumped: it was a stinker with voters, to be frank. In *Sydney Morning Herald*. City: Sydney.

Dobrowksy, P., Mannel, D., De Kwaadsteniet, M., Prozesky, H., Khan, W., & Cloete, T. (2014). Quality assessment and primary uses of harvested rainwater in Kleinmond, South Africa. *Water SA, 40*, 401–406.

Dolnicar, S. & Hurlimann, A. (2010). Desalinated versus recycled water – what does the public think? In I. C. Escobar & A. Schäfer (Eds.), *Sustainable Water for the Future: Water Recycling Versus Desalination* (pp. 375–388). Amsterdam: Elsevier B.V.

Dolnicar, S., Hurlimann, A., & Grün, B. (2014). Branding water. *Water Research, 57*, 325–338.

Dolnicar, S. & Schäfer, A. I. (2009). Desalinated versus recycled water: Public perceptions and profiles of the accepters. *Journal of Environmental Management, 90*(2), 888–900.

Domènech, L. & Saurí, D. (2011). A comparative appraisal of the use of rainwater harvesting in single and multi-family buildings of the Metropolitan Area of Barcelona (Spain): social experience, drinking water savings and economic costs. *Journal of Cleaner Production, 19*, 598–608.

Doria, M. d. (2010). Factors influencing public perception of drinking water quality. *Water Policy, 12*, 1–19.

Environment Canada. (2011). *Wise water use*. Government of Canada.

Fennell, J. O. N., & Kielbasinski, O. (2014). Water without borders. *Water Canada, 14*(6), 50–51.

Fielding, K., Gardner, J., Leviston, Z., & Price, J. (2015). Comparing public perceptions of alternative water sources for potable use: the case of rainwater, stormwater, desalinated water, and recycled water. *Water Resources Management, 29*, 4501–4518.

Fielding, K. S. & Roiko, A. H. (2014). Providing information promotes greater public support for potable recycled water. *Water Research, 61*, 86–96.

FAO (Food and Agriculture Organisation of the United Nations). (2013). Aquastat (online data base). In Countries, regions, transboundary river basins. Rome: FAO. Retrieved June 12, 2013, from http://www.fao.org/nr/water/aquastat/countries_regions/index.stm.

Freedom House. (2011). *Freedom in the world 2011: The authoritarian challenge to democracy*. Washington: Freedom House. Retrieved November 14, 2011, from http://freedomhouse.org.

Gabe, J., Trowsdale, S., & Mistry, D. (2012). Mandatory urban rainwater harvesting: learning from experience. *Water Science & Technology, 65*, 1200–1207.

Gibson, F. L., Tapsuwan, S., Walker, I., & Randrema, E. (2015). Drivers of an urban community's acceptance of a large desalination scheme for drinking water. *Journal of Hydrology, 528*, 38–44.

Grant, S. B., Fletcher, T. D., Feldman, D., Saphores, J.-D., Cook, P. L. M., Stewardson, M., . . . Hamilton, A. J. (2013). Adapting urban water systems to a changing climate: Lessons from the millennium drought in southeast Australia. *Environmental Science & Technology, 47*(19), 10727–10734.

Haddad, B. M., Rozin, P., Nemeroff, C., & Slovic, P. (2009). *The psychology of water reclamation and reuse: survey findings and research road map*. Alexandria: WateReuse Foundation.

Heyworth, J. S., Maynard, E. J., & Cunliffe, D. (1998). Who drinks what? Potable water use in South Australia. *Water, Journal of the Australian Water Association, 25*, 9–13.

Hrudey, S. E. & Hrudey, E. J. (2006). *Safe drinking water: Lessons from recent outbreaks in affluent nations*. London: IWA Publishing.

Hurlimann, A. (2008). *Community attitudes to recycled water use and urban Australian case study - part 2*. Adelaide: Cooperative Research Centre for Water Quality and Treatment.

Hurlimann, A. (2011). Household use of and satisfaction with alternative water sources in Victoria Australia. *Journal of Environmental Management, 92*, 2691–2697.

Hurlimann, A. & Dolnicar, S. (2010). When public opposition defeats alternative water projects – The case of Toowoomba Australia. *Water Research, 44*, 287–297.

Hurlimann, A. & McKay, J. (2007). Urban Australians using recycled water for domestic non-potable use – An evaluation of the attributes price, saltiness, colour and odour using conjoint analysis. *Journal of Environmental Management, 83*, 93–104.

Jeffrey, P. & Jefferson, B. (2003). Public receptivity regarding "in-house" water recycling: results from a UK survey. *Water Science and Technology: Water Supply, 3*, 109–116.

Katuwal, H. & Bohara, A. K. (2011). Coping with poor water supplies: empirical evidence from Kathmandu, Nepal. *Journal of Water and Health, 9*, 143–158.

Kimura, K., Mikami, D., & Tunamizu, N. (2007). Onsite wastewater reclamation and reuse in individual buildings in Japan. Presented at *6th IWA Specialist Conference on Wastewater Reclamation and Reuse for Sustainability*, Antwerp, Belgium.

Lawrence, D., & Haddeland, I. (2011). Uncertainty in hydrological modelling of climate change impacts in four Norwegian catchments. *Hydrology Research, 42*(6), 457–471.

Leonard, R., Mankad, A., & Alexander, K. (2015). Predicting support and likelihood of protest in relation to the use of treated stormwater with managed aquifer recharge for potable and non-potable purposes. *Journal of Cleaner Production, 92*, 248–256.

Leong, C. (2015). A quantitative investigation of narratives: recycled drinking water. *Water Policy, 17*, 831–847.

Macpherson, L. & Snyder, S. (2013). *Downstream: Context, understanding, acceptance: Effect of prior knowledge of unplanned potable reuse on the acceptance of planned potable reuse.* Alexandria: WateReuse Research Foundation.

Marks, J. (2004). Advancing community acceptance of reclaimed water. *Water Journal of the Australian Water Association, 31*, 46–51.

Marks, J. S., Martin, B., & Zadoroznyj, M. (2006). Acceptance of water recycling in Australia: National baseline data. *Water Journal of the Australian Water Association, 33*, 151–157.

McKay, J. & Hurlimann, A. C. (2003). Attitudes to reclaimed water for domestic use: Part 1 age. *Water, Journal of the Australian Water Association, 30*, 45–49.

Morimasa, T., Shuzo, N., & Masayasu, I. (2014). Quantification of the adverse effects of drought caused by water supply restrictions considering the changes in household water consumption characteristics. *Water Science & Technology: Water Supply, 14*(5), 743–750.

Mumford, L. (1989). *The city in history*. New York, NY: MJF Books.

National Research Council. (2012). *Water reuse: Potential for expanding the nation's water supply through reuse of municipal wastewater*. Washington: The National Academies Press.

Norwegian Ministry of Petroleum and Energy. (2015). *Facts 2015 - energy and water resources in Norway*. Norwegian Ministry of Petroleum and Energy.

Novelo, J. A. M. & Tapia, L. R. (2011). The growth of water demand in Mexico City and the over-exploitation of its aquifers. In Ú. O. Spring (Ed.), *Water resources in Mexico* (pp. 395–406). Heidelberg: Springer, Berlin.

OECD. (2008). *OECD Environmental Data Compendium 2006-2008: Inland Waters*. Paris: OECD. Retrieved from http://www.oecd.org/environment/indicators-modelling-outlooks/41878136.pdf

Ormerod, K. J. & Scott, C. A. (2013). Drinking wastewater: Public trust in potable reuse. *Science, Technology & Human Values, 38*, 351–373. Retrieved from http://www.oecd.org/environment/indicators-modelling-outlooks/41878136.pdf

Özdemir, S., Elliott, M., Brown, J., Nam, P. K., Hien, V. T., & Sobsey, M. D. (2011). "Rainwater harvesting practices and attitudes in the Mekong Delta of Vietnam." *Journal of Water, Sanitation and Hygiene for Development*, 171–177.

Parag, Y. & Roberts, J. T. (2009). A battle against the bottles: building, claiming, and regaining tap-water trustworthiness. *Society & Natural Resources, 22*, 625–636.

Paulhus, D. L. (1991). Measurement and control of response bias. In J. P. Robinson, P. R. Shaver, & L. S. Wrightsman (Eds.), *Measures of personality and social psychological attitudes* (pp. 17–59). San Diego, CA: Academic Press.

du Pisani, P. L. (2005). Direct reclamation of potable water at Windhoek's Goreangab reclamation plant. In S. J. Khan, M. H. Muston, & A. I. Schaefer (Eds.), *Proceedings from integrated concepts in water recycling* (pp. 193–202). Wollongong: Wollongong University.

Price, J., Fielding, K., Leviston, Z., Bishop, B., Nicol, S., Greenhill, M., & Tucker, D. (2010). *Community acceptability of the indirect potable use of purified recycled water in South-East Queensland: Final report of monitoring surveys*. Brisbane: Urban Water Security Research Alliance.

Resource Trends Inc. (2004). *Best practices for developing indirect potable reuse projects: Phase 1*. Alexandria: WateReuse Association.

Rosa, G. & Clasen, T. (2010). Estimating the scope of household water treatment in low- and medium-income countries. *American Journal of Tropical Medicine and Hygiene, 82*, 289–300.

Roseth, N. (2008). *Community views on recycled water: the impact of information*. Adelaide: Cooperative Research Centre for Water Quality and Treatment.

Rossiter, J. R., Dolnicar, S., & Grün, B. (2015). Why level-free forced choice binary measures of brand benefit beliefs work well. *International Journal of Market Research, 57*, 1–9.

Sodha, S. V., Menon, M., Trivedi, K., Ati, A., Figueroa, M. E., Ainslie, R., Wannemuehler, K., & Quick, R. (2011). Microbiologic effectiveness of boiling and safe water storage in South Sulawesi, Indonesia. *Journal of Water and Health, 9*, 577–585.

Spring, Ú. O. (2011). *Water resources in Mexico. [electronic resource] : scarcity, degradation, stress, conflicts, management, and policy*: Berlin; New York: Springer.

Sydney Water. (1996). *Community views on water reuse*. Sydney: Research Report, Sydney Water.

Theodori, G., Avalos, M., Burnett, D. B., & Veil, J. A. (2011). Public perception of desalinated produced water from oil and gas field operations: a replication. *Journal of Rural Social Sciences, 26*, 92–106.

Water Corporation. (2015). Water Sources. Retrieved 15 December, from http://www.watercorporation.com.au/water-supply-and-services/rainfall-and-dams/sources

Water Supply and Sanitation Technology Platform. (2010). *Managed aquifer recharge: Enhancing groundwater resources within and integrated water resource management*. The European Water Platform, European Commission. Retrieved March 26, 2013 from: http://www.wsstp.eu/files/WSSTPX0001/library/managedaquiferrecharge/STSManagedAquiferRecharge.pdf.

Wetterau, G., Liu, P., Chalmers, B., Richardson, T., & Boyle VanMeter, H. (2011). Optimizing RO design criteria for indirect potable reuse. *IDA Journal of Desalination and Water Reuse, 3*(4), 40–45.

World Bank. (2012). *2012 world development indicators*. Washington: The World Bank.

World Water Assessment Programme (United Nations). (2009). *Water in a changing world / World Water Assessment Programme*. Paris: UNESCO Publishing; London: Earthscan, c2009. [3rd ed.].

Yale Center for Environmental Law and Policy, Yale University Center for International Earth Science Information Network, Columbia University, World Economic Forum, Joint Research Centre, and European Commission. (2005). *2005 Environmental Sustainability Index*. Palisades, NY: NASA Socioeconomic Data and Applications Center. Retrieved 14 November 2011, from http://sedac.ciesin.columbia.edu/data/set/esi-environmental-sustainability-index-2005.

Yamagata, H., Ogoshi, M., Suzuki, Y., Ozaki, M., & Asano, T. (2002). On-site water recycling systems in Japan. Presented at *World Water Congress*, Melbourne.

Index